Mirko Daniel Garasic

Guantanamo and Other Cases of Enforced Medical Treatment

A Biopolitical Analysis

Springer

Mirko Daniel Garasic
The Edmond J. Safra Center for Ethics
Tel Aviv University
Tel Aviv
Israel

ISSN 2211-8101 ISSN 2211-811X (electronic)
SpringerBriefs in Ethics
ISBN 978-3-319-22652-1 ISBN 978-3-319-22653-8 (eBook)
DOI 10.1007/978-3-319-22653-8

Library of Congress Control Number: 2015947410

Springer Cham Heidelberg New York Dordrecht London

Printed on acid-free paper

Springer International Publishing AG Switzerland is part of Springer Science+Business Media (www.springer.com)

SpringerBriefs in Ethics

More information about this series at http://www.springer.com/series/10184

Ai miei amatissimi genitori, grazie
ai quali sono.

Preface

In the course of this book, I will foreground a number of cases in which the supposedly objective way of portraying autonomy has been in fact applied in rather different ways in relatively similar cases of enforced medical treatment occurring in the Western world. More specifically, I will focus my attention on the US, the UK and Italy.

The current approach used in bioethics, and more specifically medical ethics, gives a prominent role to the notion of autonomy when dealing with sensitive issues related to the patient's future. This central notion of autonomy was necessary for the construction of the premises of the ethical revolution that shook the Western medical world after WWII.

As a response to the atrocious medical experiments carried out by German and Japanese doctors, the Western world wanted to ensure greater protection for the patient such that he might be better able to defend himself from treatment forced upon him in the name of [pseudo] science. This was achieved via the implementation of the notion of informed consent, through which the decisional power of the patient increased drastically, the results being seen on many occasions in direct improvements to the possibility for affirmation of autonomy, freedom of choice. In recent years, however, the nature of such improvements has been called into question.

A second foundational element of the role of autonomy in Western bioethics has been one particularly important shift in medical ethics over the past 20 years: the patient–doctor relationship has moved from a paternalistic model—whereby doctors were expected and entitled by law to enforce on patients their judgement on the presumption of "knowing best"—to a new system where a patient's authority over her body is central. This qualitative change to the patient's decisional power has in itself increased the (quantitative) weight of autonomy in specific bioethical controversies. Gradually, society has modified its perception of autonomy in medical contexts, moving from a concern with the best possible option for the patient—for whom any deviation from this path only served to further negate his autonomy due to a putative lack of competence—to an increasing respect for the patient's autonomy on condition of sufficient proven competence. The growing acceptance of the patient's will as a sufficient moral justification for ensuring, or withdrawing, treatment has created a number of controversial cases in relation to the patient–doctor relationship.

Thus, as a starting point I will consider the current Anglo-American legal system—whereby the notion of respect for patient's autonomy has increasingly gained more relevance vis-à-vis the previous paternalistic approach that was dominant in the patient–doctor relationship. Paternalism had assumed that physicians were to be allowed to interfere with a person's freedom of action, a person's autonomy, on the grounds that it is for the good of the person, her liberty thus being legitimately restricted.

The growing number of debated cases of refusal of medical treatment and its denial on the grounds of impermissibility in the name of the patient's best interest has underlined the need for our attention to be refocused on the actual disappearance of paternalism from bioethical debates.

The crucial switch in power relations is characterised by an acknowledgement that—under satisfactory conditions of competence—the patient is the best judge for providing—or rejecting—informed consent over a medical procedure concerning her. Prior to this conception of "patient knows best" there existed a dogmatic view of doctors as the best judges by definition. After all, what can the patient be expected to understand about the procedure that she may or may not decide to undergo? Eventually, it was realised that doctors are as imperfect as any other professionals, and though generally more informed than their patients, doctors can fail just as every other human being engaging in any practical activity can. The *a priori* justification of valuing the nobility of having chosen a profession where an irreversible mistake can be noticed was no longer satisfactory. The more this knowledge of fallibility spread in society, the less people were willing to trust doctors, resulting in a need, gradually expanded, for an increase in the patient's decisional power, or, in more technical terms, in greater respect for her autonomy.

This reform brought innovations in the capability of the patient directly to shape her [medical] destiny. This was seen as a positive change in biomedicine but also more broadly in society, because, despite often being underestimated, the interconnection between "medical politics" and "real politics" is direct enough to allow one to influence the other in significant ways.

It is on these grounds that this work should be considered: through the analysis of the unconvincing application of the notion of autonomy in some specific cases, it is my intention to broaden our perspective on how we should evaluate an inconsistent use of this central concept.

In bioethics there exists a tendency to assess autonomy (or competence) as the function we can—or cannot—have within a given system. My idea is that we must reconceptualise our way of understanding the principle of autonomy by abandoning the mental state that puts a barrier between the sphere of bioethics—and biopolitics as a result—and the broader political scheme within which certain interpretations can be questioned. In understanding the inadequacy of such a premise in current debates, we must be ready to dissolve it progressively by the acceptance of its anachronism. As it will be shown through a number of specific cases that pertain to both bioethical and political debates, autonomy should be seen as a shared term that reinforces the connection between the two fields. This questioning our own stance on many delicate issues is a necessary means to avoid a situation in which the inconsistency present in the two spheres of justice (political and bioethical) produces such unhappiness—through the biased use of autonomy as a tool functional to power and not to individuals—that the very groundwork of the current Western society could be shaken by a violent outburst of anger towards authorities, the state and the status quo more generally. If we are to defuse this tension, we are thus required to provide a less drastic change (operated within the current system), yet we need also to realise the urgent need for innovative examination of the role granted to autonomy in Western society.

In order to reveal the central goal of the present investigation, I will attempt to cover a range of different cases exhibiting a certain commonality yet also varying along other important axes. This will allow for the gradual broadening of the reader's perspective, ultimately demonstrating the interconnection of all of the specific cases (related by the use of autonomy as their basic principle of justification) and their political contexts.

In Chap. 1 of this book, I will explore standard accepted versions of the notion of autonomy in Western contexts (particularly those of Kant and Mill), considering as well the interconnected parallel notions of competence and biopolitics. A fundamentally important aspect that should be understood from the beginning is that, when referring to the incoherence of the application of the notion of autonomy in this work, I will not aim to point out a tension between the Kantian and Millian versions, as I do not contend that one is exclusive of the other. I am aware that these two conceptions could be used in parallel without undermining the consistent application of the wider notion. My critical analysis of how autonomy is subject to contingent interpretations will instead be centred upon a malfunctioning use of its definition according to Kant's description. It is within that version of autonomy that I will foreground the incoherence to which I refer.

To give my critique a more precise and detailed frame of reference, in the subsequent three chapters I will focus my attention on four specific contexts in which the concept of autonomy (and its related sub-definition of competence) has been applied in an inconsistent and therefore questionable manner. In concluding Chap. 3 I will provocatively sustain that in future cases resembling those considered, the medication of mentally ill death-row prisoners scheduled for execution should be avoided as it would be the only way to ensure a more coherent way of applying the principles that we—as a society—claim to defend. The reasoning behind such a provocation will bring into the equation the direct relevance of politics in defining our ways of dealing with this bioethical case. Finally, Chap. 5—with its reference to Guantanamo—will make the connection between bioethics and politics even more evident.

Having realised the structural limits of the individual-centred version of autonomy that governs the bioethical and political world, as well as all the problems related to its misuse as a natural reaction preserving the very system that we live in, we have the moral duty—and we should have the political wisdom—to reshape the autonomy discourse towards a more communitarian conception that will help us deal with future cases. In order to ensure the relevance of this book in progress towards this end, however, some important premises to my work must be made.

First of all, I am aware of the fact that each of the cases considered could produce sufficient material for a book of its own. However, I urge the reader to understand that despite its intriguing appeal at theoretical level, the option of expanding the analysis further for each case would have led the book to sacrifice breadth in favour of depth given the practical constraints of this work; while focusing on a single case also represented a valid option, this would not have allowed for a wider evaluation of the role of autonomy. The payoff is the ability to consider what is common to these different (and yet sufficiently similar) situations.

Indeed, this relates to the main objective of this work: to reconnect the discussion over autonomy taking place in the field of bioethics to its political context, interrogating the current conviction that bioethical cases should be evaluated as a separated field altogether.

In this respect, I think it would important for the reader to understand how this book developed into its current form, as the research process itself has undoubtedly played a central part in the shaping of the work.

Initially, my research was centred upon Anorexia Nervosa and the debate over the acceptability of refusal of naso-gastric treatment by patients suffering from this unique mental disorder. My perception of the problem was that if we accept that autonomy is the evaluating factor upon which we should base the moral and legal permissibility of an action, all we had to do was to establish if anorexics are autonomous. As will be explained in Chap. 2, this debate is related specifically to the assessment of the presence of competence and/or capacity (the terms with which we connote autonomy in medical contexts), two related concepts whose definition is problematic in itself.

However, what became gradually more obvious to me was that, if I wanted truly to understand what made cases of Anorexia Nervosa so controversial, focusing the argument only on the assessment of competence as the way of resolving its controversial status would have provided only a temporary answer.

To grasp the depth of the issues at stake, I had to increase the challenge to case-specific analyses and move the investigation to more structural questions regarding autonomy and its role in Western bioethics. The decision to broaden the coverage of my research allowed me gradually to question my initial idea of the assessment of competence in Anorexia Nervosa as a sufficient guide to the permissibility of medical treatment (or its refusal).

Firstly, I encountered a case involving schizophrenia and capital punishment (exposed in Chap. 3) that quite drastically contrasted with the common idea of forcing medical treatment on a person to keep them alive, as in this case the notion of autonomy was used to justify enforced medical treatment to kill—albeit indirectly—the person that should supposedly have benefited from the treatment.

This strikingly different way of defining how we as a society should decide to respect autonomy provoked further questions in me. I came increasingly to doubt the absolute to which we refer when talking about autonomy, understanding that ultimately it is its interpretation (related to human beings within a predefined political structure and thus subject to power relations) that really makes a difference between being forced to stay alive and being allowed to die.

For this reason, in Chap. 4, I decided to broaden even further the spectrum of the cases, taking into account a number of comparable situations where respect for the patient's autonomy had been inconsistent and in Chap. 5 I chose to engage with perhaps the most representative and challenging case of our time in relation to the issues considered.

The need to consider these cases was vital to the strengthening of the validity of the critical analysis engaged in up to that point: after having introduced the reader to the complexity of conceptions of autonomy in Western bioethics through the contrast of enforced treatment in cases involving mental illness (and thus competence), I realised that considering contexts in which the lack of mental competence was less of a central issue might underline even more clearly the fact that the real entitlement to use one's autonomy is a function of its political acceptability.

This biopolitical strand of the book (through a reading of the similarities and differences of the cases) allowed me to point out that—contrary to the tendencies of much of the American bioethical community—bioethics cannot, and should not, be considered as a different field from the rest of philosophy. And, most importantly, we should not think that the autonomy to which we refer when debating a certain given practice or policy is not affected by the political context in which it develops, takes place and becomes a bioethical reality.

As already noted in passing, due to the structural limits of scope associated with a book, not every tangential point worthy of attention could be exhaustively explored, and for this reason I unhesitatingly acknowledge that (as in every research project) my choices as to what is relevant and what is not are open to challenge and to criticism. Given the intention of this work to link various subjects not commonly considered in such close proximity, there is at once a greater risk of undervaluing certain dimensions of the problems tackled and building in structural faults from the outset. Nonetheless, I believe that I have managed to produce a coherent and linear argument that is not jeopardised by the necessary underdevelopment of certain peripheral topics.

Some final clarifications should be made before proceeding into the main body of the book. Firstly, I have voluntarily chosen to use—even in cases of major authors—a limited number of sources from which to derive quotations and direct references. The reason behind this decision is the conviction that applying a more balanced and equal representation of the work of both the more and the less well-known philosophers used herein will prevent the reader from becoming distracted by the potential inputs that each of them might have had if moved into a more central role. My priority is rather to ensure that none of the sources applied in this context obfuscates the central argument of this work.

Hence, the use of Foucault, for example, is limited in terms of utilisation both of space and of literature, but there are two reasons for this. The space given to his work is limited because I did not want to make this book a Foucauldian one, but only to use some of his more valuable insights in support of my project of reconnecting bioethics to its philosophical roots. Foucault proved very useful and apt in this enterprise, yet I did not want the book to be absorbed by his ideas.

That is why—especially in Chap. 4—I refer mainly to one text of his. Without wanting to deprive him of his well-deserved renown nor deny the validity of his broader analysis of power relations, my intention was to treat him as all of the other authors used in the book are treated. That is, using their ideas only where they serve to develop the work towards its intended trajectory.

Lastly, I want explicitly to affirm that the interchangeability of "she/her" and "he/him/his" is also intentional. This decision might not satisfy every reader in stylistic terms, but it is the most convincing way for me to ensure that the book remains gender neutral without depriving the individuals considered within (sometimes not directly named but still existent) of their humanity.

Acknowledgments

As an evolvement of my Ph.D. thesis, this book would not have been possible without a large number of people that I want to thank. Of my professors and colleagues at the Center for Ethics and Global Politics, LUISS University I particularly appreciated the inputs received from Daniele Santoro, Aakash Singh Rathore, Mario De Caro, Valentina Gentile and Sebastiano Maffettone.

From my research period in the UK, special thanks go to Jacinta Tan, Tony Hope, Michael Parker and—of course—Charles Foster (with whom I collaborated on the Ben Garci case), who all helped me in more than just academic contexts during my stay at the Ethox Centre, in Oxford. I am also grateful to Simona Giordano for her genuine support of my initial project.

Of the people I interacted with while in Mumbai, meanwhile, I want to thank Roy Nobhojit at the Tata Institute of Social Sciences and Neha Madhiwala at the Center for Studies in Ethics and Rights for having contributed to making my time in India productive and enriching. Nicole Vincent also deserves my gratitude for her help and support in the very beginning of this work.

From my period abroad in the US, I have to thank above all Daniel Callahan and Gregory Kaebnick at the Hastings Center and David Smith, Robert Burt, Carol Pollard and Stephen Latham from the Yale Interdisciplinary Center for Bioethics, Yale University. I feel honoured to have met and had the chance to discuss my work with these great scholars and I am thankful for the interesting insights that I received from them.

My time at the Center for Human Bioethics at Monash University as an Erasmus Mundus Fellow and my year at the Hebrew University of Jerusalem as a Golda Meir Postdoctoral Fellow have not been devoted specifically to the development of this book, but nonetheless I feel like thanking a number of scholars with whom—through interesting and stimulating discussions—I have exchanged ideas that certainly affected the shaping of this text. Out of a large group of people, I want to thank Justin Oakley, Michael Selgelid, Robert Sparrow, Catherine Mills, Alberto Giubilini and Francesca Minerva from my research period in Melbourne and Daniel Sperling, Avner De-Shalit, Daniel Swartz and David Enoch from my time in Jerusalem.

Finally, I am indebted to many of my current colleagues at the Edmond J. Safra Center for Ethics at Tel Aviv University. In particular, I want to thank Yechiel Michael Barilan, Shai Lavi, Haim Hazan, David Heyd and Hagai Boas for having built such a lively and active research environment—and for having included me in it.

I also want to thank Asa Cusack, Zohar Lederman, Anna Plyushteva, Filippo Celuzza, Alison Ainley, Nadav Davidovitch, Andrea Lavazza, Jonathan Moreno and the anonymous reviewers of the manuscript for the different inputs they gave to this work.

Contents

Chapter 1
The Concepts of Autonomy, Competence and Biopolitics

1.1 Introduction

In order to enter the specificity of the debates related to the application of the notion of autonomy in current bioethical contexts, I shall first construct a more general framework within which the cases presented in the following chapters will be evaluated. Starting from a historical analysis of autonomy, I will then apply it to biomedical contexts, drawing a critical map of the current inconsistencies in its application to cases of enforced treatment (be it by not allowing withdrawal of medical treatment or by forced continuation of medical treatment depending on the circumstances), suggesting ultimately that a possible solution for properly identifying possible improvements in our approach could come from an investigation of the political meaning and value of the choices made by (medical) authorities.

1.2 Autonomy

Autonomy is a notion that has been present in philosophy since ancient times, and it results from the combination of two old Greek words: *auto* (self) and *nom[os]* (rule/law). *Autonomia*, or autonomy in English, tends to be translated into the literal rendering of "self-ruling". The original context in which it was introduced was the Greek city-state (*Polis*), Plato having coined this neologism in the course of his work *The Republic*.[1] For this reason, it seems obvious why autonomy was initially utilised only in political contexts and was not related to individual autonomy (though I am aware that some of the interpretations of Plato's ideal city have suggested that it constitutes a metaphor of a human being -resulting in a use of the notion related to a single individual- this distinction will not be investigated further here).

---

[1]Plato (1984).

M.D. Garasic, *Guantanamo and Other Cases of Enforced Medical Treatment*, SpringerBriefs in Ethics, DOI 10.1007/978-3-319-22653-8_1

Despite these social origins, the most common way of defining personal autonomy has been through the idea of self-governance: the possibility of expressing one's own personality, preferences and uniqueness without external interference, be it physical or psychological. According to two of the major Anglo-American bioethicists, Beauchamp and Childress:

> The autonomous person determines his or her course of action in accordance with a plan chosen by himself or herself. Such a person deliberates about and chooses plans and is capable of acting on the basis of such deliberations, just as a truly independent government is capable of controlling its territories and policies.[2]

A person who is judged to have limited autonomy, as in cases of Anorexia Nervosa, for example, is instead under the control of others, constrained in putting into action her own judgements by the limits placed on her by them, and thus deprived of her state of self-government. This political reading of the notion of autonomy will be analysed in greater depth later in this chapter, but first we shall focus on the term autonomy as applied to individuals in a broader sense.

Two prominent figures within the philosophical community in the Western tradition have been responsible for developments crucial to understandings of autonomy: Immanuel Kant and John Stuart Mill. The former saw autonomy as freedom of the will, the latter as freedom of action. This distinction is crucial to this work, and therefore I will now go on to explore in greater depth the respective positions of the two philosophers.

In numerous of his writings,[3] Kant argued in favour of his well-known maxim that people should always be treated as autonomous ends in themselves and never merely as means to the ends of others. One of the formulations of his Categorical Imperative affirmed that: "I should never act except in such a way that I can also will that my maxim should become a universal law."[4]

Kant's principle of respect for autonomy shows the relevance given to the value of autonomy: persons are unconditionally worthy rational agents that cannot be treated as "things" of conditional value, incapable of making decisions. In his analysis of autonomy, Kant contrasts it with heteronomy (rule by other persons or conditions). On the one hand, being autonomous means to be able to govern oneself in accordance with moral principles contained in the multiple formulations of Universal Law,[5] which could be willed to be valid for everyone. Acting heteronomously, on the other hand, can include both external and internal determinations of the will, but it does not include moral principles. One can act in accordance with what the law establishes to be "right" (not refusing medical treatment, for example) but if the resulting action does not arise from an internal autonomous acceptance of the validity of such an action as a universalisable one, it would mean that the given

[2]Beauchamp and Childress (1983, pp. 59–60).

[3]Kant (1998), (1981).

[4]Kant., I., Grounding for the Metaphysics of Morals, *ibid.*, p. 14.

[5]For example, Onora O'Neill distinguishes between the formula of universal law (FUL), the formula of the end in itself (FEI) and formula of autonomy (FA). O'Neill (2004, pp. 93–109).

behaviour would only be a forced result of the specific (political) circumstances in which the individual existed. This is a crucial distinction to which we shall return as the work evolves.

While Kant focuses on the moral dimension of autonomy, Mill is more concerned with establishing a system to address autonomy in more practical terms. In fact, he even prefers to refer to autonomy in a slightly different manner as the individuality of action and thought. In *On Liberty* Mill points out that we can accept social and political control over individuals as legitimate only if it is a necessity to prevent harm to other individuals. The principle of utility constructed by Mill stresses exactly this system of drawing the line of acceptability: every citizen is free to develop her potential according to her preferences, as long as the resulting actions do not interfere with an equivalent freedom of expression that must be granted to others. In Mill's eyes, in contrast with imposing a standardised set of policies that damages society by reducing individual productivity and creativity, what maximises the shared benefits of the community -its utility, as it were- is the promotion of autonomy, meaning promotion of one's own values and priorities. In this line of thought Mill affirms that only those persons "without character" keep on being influenced and controlled by authorities such as the state, parents or the church. In fact, it is only the person with true character that has a genuine individuality. In other words, even if perhaps in a less Nietzschean way than I am presenting him, Mill suggests that only those capable of detaching themselves from the predefined notions of good and bad are in truth autonomous, and in some sense, worthy of true respect and consideration. This certainly is not only my interpretation but is rather a widely accepted view. On this issue, Beauchamp and Childress write:

> 'Firmness and self-control' as well as 'choosing a plan of life' are declared by Mill as essential to a proper framing of one's character. 'The government of a strong will' he takes to be essential to this goal.[6,7]

Mill's position is interesting for two reasons: one, it presents a dissident citizen as the most likely candidate to be autonomous in political terms. Two, the above affirmations become even more striking in relation to a specific case that will be considered more in depth in the next chapter, namely that of Anorexia Nervosa. In this context, indeed, it should be stressed that the vast majority of people suffering from this mental disorder are indeed strong personalities capable of affirming themselves in many different contexts (school, work, gym) through the application of their "strong will, firmness and self-control".

It appears clear, then, that Mill and Kant had different intentions when analysing the concept of autonomy. The former gave relevance to the personal point of view in his account of autonomy and self-determination. He argued in favour of the respect of one's individuality as the primary element in ensuring a morally acceptable form of autonomy. The latter focused on the moral dimension of

---

[6]Beauchamp and Childress (1983, p. 61).

[7]Mill (1974, pp. 136–138, 184–189).

autonomy: the idea of following a moral law in which the notion of self-determination is acceptable only within a pre-set framework of morally valid principles. The substantial difference between the positions is that, for Kant, purely individual actions are outside the moral order.

Bruce Jennings suggests that perhaps a way of better understanding of what we mean by autonomy in a contemporary context would be to use the term *autonomy* to refer to the definition given by Kant, as opposed to *liberty* when referring to Mill.[8] In addition, Jennings points out that frequently in the field of bioethics what we mean by autonomy is not the Kantian version of the notion but something much closer to Mill's idea of liberty, as subsequently expanded by Isaiah Berlin into the distinction between negative and positive liberty (the common bioethical version of autonomy falling within the former category).[9] This schematisation of the two variants would lead us to see that such a distinction in the current way of understanding autonomy is not as neat as it might initially seem. It is no surprise, in fact, that when looking more closely at how we come to define autonomy in practical terms, a synbook of the two notions appears.

I agree with Jennings' affirmation that we experience a form of autonomy that bears greater resemblance to Mill's *liberty* than to Kant's *autonomy*: in most cases in biomedicine we consider the individual values as the ultimate way of affirming one's autonomy, and we therefore consider it morally acceptable. Without entering into the details of the debate and whatever position one might have on the topic, it is undeniable that the famous argument of Judith Jarvis Thomson in defence of abortion represents an example of the acceptance of the precedence of the individual in both the academic and the legal systems.[10] In some cases, however, the acceptability of a decision goes back to the authorities on the presumption that the patient could not possibly wish for treatment to take place or be withheld. As this work develops, I will consider some political aspects of the complex concept of autonomy, but before proceeding with an analysis of autonomy in bioethical contexts, I should point out that the reason for choosing to highlight -albeit in a limited way- such dimensions of autonomy is based on the conviction that an analysis of the political implications is necessary to any understanding of specific applications of autonomy in enforced medical treatment.

In relation to what has been explained above, it is important to underline that the nature of the justification for this switch back to paternalism -and with it the notion of "knowing best"- requires that moral dimension of autonomy we already referred to: we are entitled to force someone into doing something because we are already aware

[8]Jennings (2007, pp. 75–76).

[9]Berlin (2002, pp. 118–172).

[10]In her famous defence of abortion, Thomson presents us with a thought experiment: if we would wake up (after having been kidnapped by the Society of Music Lovers) attached to a famous violinist whose life would depend on us, we could choose to go *beyond* our obligations and stay attached to him for 9 months to save his life, but if we did not, we could not be labelled to be selfish or inhumane: 9 months is a very long time and it represents a level of exceptional commitment that one can refuse to undergo. Thomson (1986, pp. 37–56).

of the existence of a pre-set framework of morally valid principles that we try to help the "enforced" to follow. In other words, we would presume that an optimal solution to any situation can be achieved through the application of reason to the specific case. However, in this way, we would be applying a Kantian approach rather than a Millian one.[11] It is for this reason that O'Neill's attempt to reframe our manner of adapting Kant to bioethics is of crucial importance, and it deserves attention. Her book *Autonomy and Trust in Bioethics*[12] lays the foundations of a conception of autonomy that sees the autonomous agent as one obliged to make her choices within a predefined moral set of normative values. As rightly pointed out by Anne-Marie Slowther,[13] even if O'Neill has a personal preference in defining the principle of duty or obligation as the decisive factor in establishing that a choice is autonomous, other value frameworks are possible. The peculiar aspect of this approach to the notion of autonomy is the relevance given to responsibility. The responsibility of the agent taking the decision differs greatly from the previous definition of autonomy. In a very Kantian fashion, O'Neill suggests that the means of reaching a truly autonomous judgement in a given situation are available, but it is up to the individual in question to make an effort in following -and even interpreting- the already present moral framework that can ensure autonomy. Gordon Stirrat and Robin Gill[14] went further in O'Neill's direction in reaching the conclusion that a principled autonomy in the case of patients would require the patient to choose her medical treatment responsibly and in consideration of others. On the one hand, this suggestion aggravates the differences with the previous model of autonomy, as for those not sharing an O'Neillian formulation of principled autonomy, a "requirement" of this kind could easily be seen as an external constraint limiting the patient's liberty (which is to say, autonomy). On the other hand, this application of Kant seems to be much less deontological than classical interpretations.

However, Jennings argues that O'Neill's attempt at reviving the use of Kantian ethics in bioethical debate is bound to fail since there is a clear distinction between academic philosophy and bioethics (at least in the USA, he affirms) and defining autonomy in a satisfactory way for both would be impossible. The reason for this impossibility lies in the fact that Jennings believes that certain Western values are so rooted in our society that it would be unrealistic to presume that people would be able to detach themselves from notions such as individualism or self-expression.

In other words, the way in which Western society has referred to autonomy in recent decades has generated a subconscious popular adaptation to what is acceptable and what is not based on the version of autonomy that has been

[11]Surely the distinction between the Kantian and Millian versions of autonomy provided in this chapter is far from satisfactory if perceived as an attempt to give a conclusive portrayal of the ideas of the authors. This, however, is obviously not my intention. My aim is to give a general version of the two approaches such that the reader can understand the reasons behind my inclination towards a Kantian analysis as a way to solve the controversial cases considered in this book.

[12]O'Neill (2002).

[13]Slowther (2007, p. 173).

[14]Stirrat (2005, pp. 127–130).

established as the only acceptable one. This applies to bioethics as much as to other contexts, and for this reason when I think of the political significance of autonomy in the course of history I have to disagree with Jennings' position. His attempt to limit what I believe to be a very interesting contribution made by O'Neill seems to suggest that, in the course of history, initiatives that intended to reveal the limits of the prevailing view at the time could never have taken place owing to a lack of initial support.

In addition to O'Neill's account, in recent times there have been many and varied attempts to define autonomy in the contemporary medical ethics literature, each one of which highlights particular aspects of the concept. I will schematize below the most relevant ones for the present work.

1.2.1 Autonomy as Freedom to Have One's Will Respected

The representative of this category is the already mentioned Berlin.[15] In fact, the libertarian view of autonomy that is understood as the freedom to choose between different options without external restrictions or obligations seems to correspond significantly to Berlin's concept of negative freedom, which -we should not forget-evolved out of Mill's concept of liberty and thus came to be defined as "libertarian". In such cases, respect for autonomy would be limited to the acceptance of the patient's will without any evaluation of the validity of such a choice. Enforced treatment thus could not be justified where the patient is considered to be sufficiently competent to give -or deny- her permission for the application of a particular healthcare procedure. It is very important (but often overlooked) that implementing this method of dealing with the issue of autonomy must bring the clinician's role into the equation. In other words, the acceptance of respect for the patient's autonomy does not, and should not, assume an automatic responsive duty on the part of doctors involved in the patient's treatment.

1.2.2 Autonomy as Substantive-Procedural Conception

In her book *Understanding Eating Disorders* Simona Giordano underlines further the link made between autonomy and practical rationality, pointing out its limits especially in relation to mentally ill patients. This group is especially at risk of not being eligible to express their will due to the presence in wider society of an embedded acceptance of notions described by John Rawls (ideal rationality)[16] and Danny Scoccia (social acceptability).[17] However, as John Harris points out,

[15]Berlin, I., Op. Cit.

[16]Rawls (1999, pp. 248–250).

[17]Scoccia (1989–1990, pp. 318–334).

in situations such as that of anorexic patients, not considered "genuinely autonomous", we would end up tolerating the paradoxical situation of claiming to respect them by not respecting what they really want.[18] Clearly, this premise of "genuine autonomy" risks establishing a biased approach to what is justifiable. Giordano writes: "a substantive conception of autonomy, in fact, leads to the justification of an authoritarian attitude towards the patient and disregard for patient autonomy."[19] An alternative to this controversial conception -which will later be analysed in greater depth and in relation to its political implications- is a procedural (or formal) conception of autonomy, and it is this that constitutes the legal approach to decision-making capacity in the UK, as defended by numerous liberal philosophers.[20] The key aspect is that in this latter conception, decision-making capacity is not dependent on the status[21] of the patient but is instead a decision-relative concept.[22]

1.2.3 Autonomy as Consistency with Past Decisions

In his *Life's Dominion*, Ronald Dworkin affirms that a key aspect of defining a choice as autonomous is the consideration of its consistency with past choices made by the same individual. The centrality of personal integrity, or identity, is what is most important in this model of autonomy. Respecting one's autonomy should always take into account the need on the part of the authorities to ensure that individuals -where established to be competent- be allowed the chance to live their lives in accordance with their "distinctive sense of their own character."[23] A very important development of this view was made by George Agich,[24] who, still giving major importance to the role of one's identity in assigning the level of respect for one's autonomy, expanded the entitlement to affirm an individual's choice to third parties sufficiently capable of representing (in Dworkian terminology) the individual's character. To give a practical example, the surrogate decision-maker of a patient in a vegetative state should be entitled to decide to end artificial feeding as long as she would be able to demonstrate that this decision would be in line with the values expressed by the patient over the course of her life.[25] Speculations of this

[18]Harris (1994, p. 194).

[19]Giordano (2005, p. 48).

[20]See for example: Engelhardt (2003).

[21]McHale and Fox (1997, pp. 280–281).

[22]Harris (1994), Op. Cit, Chap. 10.

[23]Dworkin (1993, pp. 222–229).

[24]Agich (2003).

[25]I am aware that the legal entitlements of such a practice vary greatly between states and sometimes even regions, and as such I will not refer to any particular legal system here, but only to the a priori condition that any such system would have to guarantee.

kind have brought us morally to justify the institution of the biological will; this is particularly relevant to our case as it leaves room for the possibility of combining respect for the patient's autonomy with the pursuit of her best interest.

1.2.4 Autonomy as Capacity to Choose Validly

A final contrasting way of defining autonomy places the emphasis not on the values of the patient as in the conception outlined above, but rather on the decision-making process. In order to establish the level of autonomy thus, we need to ensure that the patient is capable of processing the information given, reflecting on it and reaching a "reasonable" conclusion. What has to be established, in other words, is whether the patient is competent or not. This approach has produced legislation such as the Mental Capacity Act 2005[26] and the more recent Mental Health Act 2007 in the UK[27] which stipulate assessment of the patient's level of "proper" understanding of a given situation. Some similar models even suggest the necessity for critical reflection,[28] but a deeper look at each of these models makes evident the enormous dependence of an individual's practical possibility of exercising autonomy on the method of competence assessment used by the authority. This contrast between authority and autonomy, as well as the varied means of assessing the competence of patients suffering from different forms of mental impairment, are crucial aspects of this way of understanding autonomy. Before moving the investigation onto the intersection of autonomy and [bio]politics (with all its resulting controversies), the next section will expand on the role granted to autonomy in the West.

1.3 Respect for Autonomy

Since its first edition, Beauchamp and Childress' *Principles of Biomedical Ethics* has had an enormous impact on both the philosophical and medical worlds. Their most influential idea has been that in the vast majority of problematic cases in the sphere of medical ethics we are obliged "only" to choose the most appropriate option for a given situation amongst four key principles proposed by the authors: autonomy, nonmalificence, beneficence and justice. These principles have proved to be a relatively effective way of solving delicate controversies.

---

[26]Mental Capacity Act, 2005, available at: www.opsi.gov.uk/acts/acts2005/ukpga_20050009_en_1 [accessed on 4 January 2015].

[27]Mental Health Act 2007, available at: http://www.legislation.gov.uk/ukpga/2007/12/contents [accessed on 4 January 2015].

[28]See, amongst others: Dworkin (1998).

Perhaps the simplicity of this solution has been one of the main reasons for which some critics have rejected *principled ethics* as an unsatisfactory solution.[29] Without questioning the validity of the formulation in such depth, I shall focus on another aspect that often appears to limit the utility of the four principles approach: the uneven importance granted to each of them. In fact, it could hardly be denied that of the four principles autonomy remains the most relevant in current bioethical evaluations. As a result, when two or more of the principles clash, autonomy is always preferred. As Raanan Gillon[30] put it, "autonomy is first among equals". But why is that so? What is so special about autonomy?

In fact, the very affirmation of considering autonomy to be the "queen of virtues", is highly questionable in cases of refusal of treatment, especially in cases where the autonomy of the patient is considered to be affected in a way that does not allow her to be competent.

Using the "four principles approach" the patient is granted by the principle of autonomy the right to an informed choice about medical treatment as long as she is established to be of competence sufficient to decision-making without coercion or under the influence of others.

As rightly pointed out by Karen Faith, however,[31] there can surely be cases where the principle of autonomy represented by a refusal of treatment will conflict with the principle of beneficence, as this latter principle requires the physician to act in the best interest of the patient. Clearly, this is not a notion that differs much from the supposedly outmoded alternative of paternalism.

The counterargument to such a critique would affirm that the key aspect that makes the two approaches different lies in the fact that while paternalism affirms a consistent superiority of the medical judgement over the patient's -including when she is competent- the principle of beneficence undermines only temporarily the will of the patient until the true autonomy of the patient is rescued. The idea of justifying involuntary treatment on these grounds has led Marian Verkerk[32] to defend the concept of "compassionate interference". Clearly, this idea presumes that there is one "best" way of dealing with a certain situation and that it can be achieved and understood through the use of reason. This is a very Kantian way of understanding autonomy, and as such it does not follow other interpretations that the current Anglo-American system utilises in other cases, including medical ones.

In other words, we decide to restrict a patient's autonomy as long as she is not capable of reaching, or deciding not to reach, the only correct answer that she can provide and that we will accept as the valid answer of a truly autonomous agent. This, however, seems to me neither an extension of Mill's nor of Berlin's definition of autonomy, but rather appears to be inspired by Kant's. A version of autonomy that sees:

---

[29]See, for example: Harris (2003, pp. 281–286).

[30]Gillon (2003, pp. 307–312).

[31]Faith (2015).

[32]Verkerk (1999, pp. 358–368).

> the rational being of legislating universal laws, so that he is fit to be a member in a possible kingdom of ends, for which his own nature has already determined him as an end in himself and therefore as a legislator in the kingdom of ends. Thereby he is free as regards all laws of nature, and he obeys only those laws which he gives to himself. Accordingly, his maxims can belong to a universal legislation to which he at the same time subjects himself.[33]

This understanding of the true nature of autonomy in contemporary biomedical contexts will give rise to an additional question related to those two mentioned above: could autonomy have been given such a privileged position because it has functioned as a perfect deterrent for people not to question (medical) authorities? In the next section, I will seek to provide an answer.

1.4 Competence

In this section, I will interrogate a particular aspect of enforced treatment: can we convincingly affirm that the assessment of autonomy in relatively similar cases is only based on medical grounds?

To answer this question, we need first to understand the dynamics involved in choosing to accept or not the possibility that an individual can refuse medical treatment. There are two issues related to the idea that refusal of treatment should not be morally -or legally- permissible: first, an autonomous agent would never rationally choose to give up her life if free not to; second, as a consequence of the first assumption, everyone falling outside of this category should be deemed *incompetent* to take decisions regarding their own life by the *competent* authorities for the very reason that their desire to make an irrational decision reveals their deficient autonomy. In other words, the patient is not allowed to make a choice regarding her own life due to a perceived lack of autonomy, or, we might say, because authorities opt not to recognise the presence of autonomy in her. If this is the case, however, the commonly held idea of preserving the medical interest of the patient seems instead to cede ground to a conception allowing the citizen to pursue her own best interest. In other words, the acceptability of what is permissible would shift from a medical to a political dimension. To answer the initial question then, we can affirm that the assessment of autonomy in relatively similar cases is *not* only based on medical grounds, as it results from the a priori decision that the authorities in charge can or cannot choose to consider the choice of the patient as acceptable.

I will now consider the first "dogma" that considers refusal of treatment unacceptable as disrespectful of the patient's autonomy: suicide is irrational.[34]

---

[33]Kant, I., Grounding for the Metaphysics of Morals, Op.Cit., p. 41.

[34]For a matter of convenience, I will use only the term suicide in a broad sense without listing refusal of treatment or euthanasia every time.

First of all, it is important to underline that, in the past, not all societies have regarded euthanasia or even suicide as something immoral in the way that most of contemporary societies around the globe do.[35] For instance Euripides wrote:

> I hate the men who would prolong their lives
> By foods and drinks and charms of magic art
> Perverting nature's course to keep of death
> They ought, when they no longer serve the land
> To quit this life, and clear the way for youth.[36]

This quotation clearly shows that in ancient Greece the perception of euthanasia or suicide was very different from ours. The individual is expected to end her life once no longer able to help the community in any other way. The ancient Greek culture is not the only one in which the elderly were expected to understand when it was their time: "Eskimo, American Indian, and some traditional Japanese cultures have practised voluntary abandonment of the elderly, a practice closely related to suicide."[37]

This awareness should lead us to understand that to see suicide as irrational might pertain in most cases but not in all of them, because if to establish the rationality of an action we need to look at its social acceptance and effects,[38] then it is our current society that pre-establishes the limits of our autonomy. I want to underline that the mere fact that our society has reached a level of scientific development to sustain lives in even the most extreme cases does not constitute a valid ethical justification for doing so unconditionally; it is not a priori irrational not to want to use these artificial means to sustain life.[39]

In this respect, John Keown should be considered when he writes:

> Despite the major advances in medicine and palliative care witnessed by the last century, many patients, even in affluent western nations, still die in pain and distress. Some entreat their doctors to put an end to their suffering either by killing them or by helping them to kill themselves.[40]

Hume points out in "On Suicide"[41] that one of the main critiques mobilised against the acceptability of suicide in our society is based on the assumption that it is an action against God. He argues that it is inconceivable not to accept suicide as an act

---

[35]This approach has been gradually challenged in recent decades, with Switzerland, Oregon and Belgium representing -perhaps above all- countries where euthanasia is perceived more and more as a morally sound option.

[36]Euripides (1928, p. 153).

[37]Battin (1994, p. 191).

[38]See, for example: [For an act to be considered rational] "the state of affairs promoted by that choice or act must be worth promoting. That is, it must promote some objectively valuable state such as well being, achievement, knowledge, justice and so on." Savulescu (1999, pp. 405–413).

[39]This position might be gradually more accepted in Western societies (especially within bioethical circles), but I want nonetheless to stress it for sake of clarity.

[40]Keown (2002, p. 1).

[41]Hume (1986, pp. 22–23).

in accordance with divine law, for God would not have given us the possibility to perpetrate the act if he did not want us to do so. Other perspectives on suicide, such as that of Aristotle,[42] see suicide as an act of cowardice doubly unacceptable as much for its vicious meaning at an individual level as for its negative political impact on society: that is, a person should not commit suicide as this would damage the common spirit of the community. Adapting this thought to cases of Anorexia Nervosa, it could be argued that refusal of treatment cannot be tolerated by our society because this action -as provocatively suggested in the previous section-could destabilise the whole moral system that is currently in use, putting the political establishment at risk as a result.

In this respect, it is interesting to note the Brad Hooker's analysis in his "Rule-utilitarianism and Euthanasia"[43] when evaluating our contemporary situation: due to scientific developments we are more capable than ever of prolonging life, but this possibility raises a crucial question: is it not reasonable to say that in certain cases it is both more moral and rational to be willing to die quickly? After all, it is hard to disagree with John Scally when he writes: "Although it is cruel not to attempt to sustain life, it may be equally cruel to extend care unconditionally."[44]

This acceptance, then, calls into question what real value decision-making authorities responsible for patients -namely healthcare professionals- are trying to preserve at all costs if not life. Might it instead be power?

In order to be able to answer this question (and to explain the second "dogma" at stake in cases of enforced medical treatment), the next section will examine a number of more conventionally political ways of defining autonomy and their biopolitical implementations.

1.5 Biopolitics

Joel Feinberg writes: "I am autonomous if I rule me, and no one else rules I."[45]

This is certainly a definition that creates more difficulties in establishing who is entitled to claim to be defending the autonomy of the patient if not herself. An even more unorthodox reading of autonomy is the one given by Robert Paul Wolff:

> As Kant argued, moral autonomy is a combination of freedom and responsibility; it is a submission to laws that one has made for oneself. The autonomous man insofar as he is autonomous, is not subject to the will of another.[46]

---

[42]Aristotle (1976, p. 130).

[43]Hooker (1997, p. 42).

[44]Scally (1995, p. 32).

[45]Feinberg (1972, p. 161).

[46]Wolff (1970, p. 14).

In this respect, Wolff[47] gives us a suggestive way of reading Kant: even if he explicitly condemns it in some passages, Kant's philosophy does not in fact clash with the idea of anarchy. In relation to this interpretation, it is interesting to consider Morris Ginsberg's[48] explanation of Kant's clear claim not to have a duty to interfere directly in the lives of others, as long as we live rightly and in accordance with virtue and without undermining the possibility for others to do the same in their own, autonomous way. Again, this interpretation seems to suggest some problems in justifying enforced treatment.

There is one crucial aspect in understanding Kant's philosophy that deserves brief consideration: autonomy contrasts with heteronomy, and the latter is the ultimate moving principle of law enforcement within a state; without a state forcing us to do something, we would not. If we were all to act spontaneously in a moral way without caring for the legal consequences of our actions, there would be no need for the state, and everyone would be acting autonomously. In following instead what is established by the state as legal, we are not acting autonomously if we do so in order to avoid breaking the law. Obviously, this consideration also has direct implications for the level of culpability of an individual within a given legal system; these implications will be more fully addressed in Chap. 3.

What is important to notice in this instance, however, is that instead of focusing on the maximal preservation of the patient's autonomy, in Western medical ethics the leading principle since antiquity has been another: the Hippocratic Oath.[49] Under this oath there is no reference to any need to involve the patient in the decision-making process, nor, as an obvious consequence, is there any mention of the principle of autonomy. Instead the doctor is required to use her skills and abilities to benefit the patient and prevent her suffering and harm.

Clearly, this view on the matter can only make sense if we presume that the doctors in question know "the right answer". In fact, this axiom is so strongly taken for granted that we accept the complete non-involvement of the patient in the decision-making process. The justification for this decision assumes that even were she to be sufficiently competent and informed regarding her situation, her actual informed consent would not alter the value of a procedure's justification, or indeed its rejection.

On the same issue, an important factor pointed out by Tuija Takala must be borne in mind:

> Although in most cases it is true that medical professionals hold superior knowledge in terms of what would be medically best for the patient, many decisions taken in the modern health care setting are not only about the medical good, and the medical good is not the only good that people are after. Most decisions are also valuable judgements about what

---

[47]See also Wolff's more extreme claims in his article (1969), p. 608, where he affirms that: "obedience is heteronomy [sic]. The autonomous man is of necessity an anarchist".

[48]Ginsberg (1965, p. 80).

[49]See, amongst others: Edelstein (1943). Available at: www.pbs.org/wgbh/nova/doctors/oath_classical.html.

> people see as valuable to them in their current situations. And in these decisions the doctor holds no special expertise.[50]

It seems clear that this way of giving relevance to the values of the individual rather than to some external definition of value fits well with some of the previous characterisations of autonomy, particularly that of Ronald Dworkin. However, of greater relevance to the current investigation is the realisation that -in cases of Anorexia Nervosa, for example- combining the above quotation with the Giordano's acknowledgement[51] that the anorexic's "defective condition" is part of who they are[52] should lead us to question the justifiability of paternalistic intervention in their treatment even where we consider Anorexia Nervosa to be a mental illness.

In this regard, we will now focus on the role and use of mental illness (and the related notion of putatively lacking competence) as a decisive justificatory instrument in cases of enforced treatment. In order to produce a relevant understanding of mental illness that will prove useful for this part of the work, I will take into account Thomas Szasz' position on the power dynamics of mental illness, psychiatry and politics.

In line with the approach of this chapter, Szasz wants us to focus on a very important aspect of psychiatry: we should understand its historical path, and, more importantly, we should be aware of its political significance. His argument is based on the assumption that mental illness does not really exist but is constructed by our society in order to explain (and justify) the differences in dealing with similar cases that would otherwise be impossible to understand. What is peculiar about this scholar is his extreme tendency to criticise what he might be expected to support in the first place: the positive role of psychiatry in our society. His argument is so extreme that in his "Involuntary Mental Hospitalization: A Crime against Humanity" he compares the incarceration of mentally ill people to slavery, writing:

> The practice of 'sane' men incarcerating their 'insane' fellow men in 'mental hospitals' can be compared to that of white men enslaving black men. In short, I consider commitment a crime against humanity.[53]

According to Szasz, "mental illness" is a metaphor, as its definition implies no connection with any sort of mental disease or disorder. Rather, such a term has had an instrumental value in deceiving us, preventing us from understanding that the acceptance of such a definition as valid is useful from a social point of view rather

---

[50]Takala (2007, p. 228).

[51]Giordano, S., Op. Cit, p. 230.

[52]In a study on Eating Disorders published by Jacinta Tan, Tony Hope and Anne Stewart, one anorexic patient replied as follows to the question "would you make the illness magically disappear if you could?": A-"Everything. My personality would be different. It's been, I know it's been such a big part of me, and I don't think you can ever get rid of it, or the feelings, you always have a bit in you." Tan et al. (2003, pp. 533–548). See also in the same journal: Tan (2003).

[53]Szasz (1998, p. 299).

than from a medical or a therapeutic one. It follows quite obviously that if "mental illness" is not pathological, there cannot be a medical justification for authorities to attempt to protect us from such non-existent diseases. In addition, the level of uncertainty over the character of mental illness (defined by highly subjective standards for assessing the severity of the illness) is undoubtedly based in large part on the specific interaction between the patient and the psychiatrist. Such inconsistency of evaluation undermines the actual relevance of the medical role in these cases, and according to Szasz, shows once again that the use of this justification is based on our society's need to obscure its real intentions (i.e. eliminate those that do not fit within the established framework) by providing us with reasons based on "scientific facts".

Generally speaking, in Western societies it is believed and accepted that the individual "owns" her body and personality,[54] therefore the physician can only be allowed to take action in cases where the patient consents. Szasz cites[55] the explicit affirmation of John Stuart Mill: "each person is the proper guardian of his own health, whether bodily, or mental and spiritual" affirming that, obviously, "commitment is incompatible with this moral principle."[56] Therefore, there is no moral justification for hospitalising an individual against her will: neither with the intention of helping them nor with the intention of insulating others from a potential danger. Szasz' suggestion is that our society should simply deal with any arising situation in accordance with its sphere of competence. In other words, we should apply different reactions to different situations regardless of the prejudicial impression that we might have of the individual involved in a specific case; a murderer should be imprisoned, while a non-violent anti-social individual should be "punished" with moral sanctions such as social ostracism, for instance.

The practicality and validity of Szasz' approach have been and are open to question, but for the present work, the most crucial element of Szasz' whole analytical project concerning mental illness is his historical interpretation of the construction of its definition. His suggestion is that we should first of all analyse certain historical facts from a point of view distinct from that preferred by society. The standard view of mental hospitals is that they help those who are inside them and that it has always been so. The only change in mental hospitals is that our scientific progress has managed to improve their level of success in curing, or at least reducing, the negative effects of some, if not all, mental illnesses. Szasz aims to convince us that this is not the real evolutionary path that "mental illness" has taken. Nor is this the way mental hospitals developed. Szasz, like Michel Foucault[57] before him, uses the example of Paris in the seventeenth century, underlining that at that time it was not even necessary for the authorities to justify the incarceration of certain members of the community. It was not necessary for such individuals to be

---

[54]Szasz (1960, pp. 332–336).
[55]Szasz (1998), Op.Cit., p. 301.
[56]Mill (1955, p. 18.)
[57]Foucault (1991, p. 8).

defined as mentally ill; all that mattered was that they were a worry for the "respectable" community and the easiest way to eliminate the "problem" was to confine them to the Hôpital Général. The position that Szasz holds purports to show that evaluations of who is mentally ill do not differ significantly from those of the past, but of course the real intention of insulating the "good" people from the "bad" now needs to be better camouflaged. In conclusion, he is arguing that people are:

> committed to mental hospitals neither because they are 'dangerous' nor because they are 'mentally ill', but rather because they are society's scapegoats, whose persecution is justified by psychiatric propaganda and rhetoric.[58]

Again, Foucault agrees with this view on many occasions, not least when considering the Soviet Union.[59]

Clearly then, there are a number of reasons why respect for the autonomous decision of refusing medical treatment is currently minimal. One of them surely relates to the loss of power on the part of medical authorities which would result from this decision. Another might be the negative influences that such a change would have in biomedical contexts, but also in wider society. This justification, however, is a political rather than an ethical one, and my dissatisfaction with the current situation lies in the fact that such an approach should be clearly stated rather than camouflaged by certain laws that claim to defend individual rights and the patient's autonomy.

One does not have to necessarily subscribe to Szasz' extensive and structural critique of the approach that the West has had (and perhaps still has in a sense) towards mental illness. Yet, his questioning attitude is useful to the analysis carried out here. To substantiate such a political reading of the way Western society has dealt and still deals with cases of enforced treatment, the next chapter will bring specific examples to the fore in order to give some of the theoretical claims made in this chapter a more contextualised dimension.

1.6 Conclusion

This first chapter has shed more light on a topic which is itself plagued by ambiguity. The need for changes in our approach to delicate matters concerning autonomy (be it that of the patient or of the doctor) is inalienable from the times we live in. The level of uncertainty regarding how best to judge and legislate for the ever-increasing number of controversial cases in bioethics is so high that debates on these issues are no longer limited to academic contexts. On the contrary, it is more and more common to find articles or TV programmes attempting to address or explain problematic cases, in which, as has been pointed out throughout the chapter,

58Szasz (1998), Op.Cit., p. 303.

59Foucault (1988, pp. 180–183).

the notion of autonomy is frequently central. This social reaction is symptomatic of a period of "reassessment of values" which calls for the attention of experts in the field of biomedical ethics, above all academics. Our response should certainly not bend to what the masses want to hear in order to preserve power, but should instead attempt to understand the core of the problem: namely, we lack a just approach when dealing with subtly yet importantly differentiated cases concerning the freedom of the individual. Especially in instances of *pro-life* versus *pro-choice* conflicts, or, as explained at the outset, where respect for autonomy and medical paternalism are juxtaposed. In the course of this chapter, I have explained the evolution of the notion of autonomy from antiquity to the present day, highlighting a number of modern interpretations of the notion and its sub-definition of competence. Subsequently, I argued that over the past twenty years the increasing importance in biomedical decisions of individual autonomy has served as a tool towards the preservation of the *status quo*. The final section of the chapter reflected on the legitimacy of affirming mental illness as an objective medical assessment, since it is also subject to the same dynamics of biopower and control. The proceeding chapters will shift our analysis to a more practical level, beginning with a more thorough examination of controversial cases of Anorexia Nervosa.

References

Agich, G. 2003. *Dependence and autonomy in the old age: An ethical framework for long-term care*. Cambridge: Cambridge University Press.

Aristotle. 1976. *Nicomechean ethics*, Trans. J.A.K. Thomson. London: Penguin Books Ltd.

Battin, M.P. 1994. *The least worst death*. Oxford: Oxford University Press.

Beauchamp, T., L., and J.F. Childress. 1983. *Principles of biomedical ethics*, 2nd edn. Oxford: Oxford University Press.

Berlin, I. 2002. *Two concepts of Liberty*, in Four Essays on Liberty. Oxford: Oxford University Press.

Dworkin, R. 1993. *Life's dominion*. London: Harper Collins.

Dworkin, G. 1998. *The theory and practice of autonomy*. Cambridge: Cambridge University Press.

Edelstein, L. 1943. *From the hippocratic oath: text, translation, and interpretation*. Baltimore: Johns Hopkins Press.

Engelhardt, T. 1996. *The foundation of bioethics*, 2nd edn. Oxford: Oxford University Press. and The many faces of autonomy. *Health Care Annual* 9.

Euripides. 1928. *Suppliants* 1109, as quoted by [pseudo-] Plutarch, *A Letter of Condolence to Apollonius*, 110C. Trans. F.C. Babbitt. Plutarch's Moralia, vol. 2. Cambridge: Harvard University Press.

Faith, K.E. 2015. *Addressing issues of autonomy and beneficence in the treatment of eating disorders*. http://www.nedic.ca/knowthefacts/documents/Addressingissuesofautonomyandbeneficence.pdf.

Feinberg, J. 1972. *The idea of a free man*, in Education and the Development of Reason, ed. R.F. Dearden. Routledge and Kegan Paul.

Foucault, M. 1988. Politics, philosophy, culture: Interviews and other writings, 1977-1984/ translated [from the French] by A., Sharidan and others, ed. L.D. Kritzman. New York: Routeledge.

Foucault, M. 1991. *The Foucault reader*, ed. Rabinow, P. Harmondsworth: Penguin.

Gillon, R. 2003. Ethics needs principles -four can encompass the rest- and respect for autonomy should be 'first among equals'. *Journal of Medical Ethics* 29.
Ginsberg, M. 1965. *On justice in society.* London: Penguin Books Ltd.
Giordano, S. 2005. *Understanding eating disorders.* Oxford: Oxford University Press, 2005.
Harris, J. 1994. *The value of life.* London: Routledge.
Harris, J. 2003. In praise of unprincipled ethics. *Journal of Medical Ethics* 29.
Hooker, B. 1997. *Rule-utilitarianism and Euthanasia*, in Ethics in Practice, ed. H. LaFollette. Oxford: Blackwell Publishers.
Hume, D. 1986. *Of Suicide*, in Applied Ethics, ed. P. Singer. Oxford: Oxford University Press.
Jennings, B. 2007. *Autonomy*, in The Oxford Handbook of Bioethics, ed. B. Steinbock. Oxford: Oxford University Press.
Kant, I. 1981. *Grounding for the metaphysics of morals*, Trans. J.W. Ellington. Indianapolis: Hackett Publishing Company.
Kant, I. 1998. *Critique of pure reason*, Trans. P. Guyer, and A. Wood. Cambridge: Cambridge University Press.
Keown, J. 2002. *Euthanasia, ethics and public policy.* Cambridge: Cambridge University Press.
McHale, J., and M. Fox. 1997. *Health care law.* London: Maxwell.
Mill, J.S. 1955. *On liberty.* Chicago: Regnery.
Mill, J.S. 1974. *Utilitarianism, on liberty, and essay on Mill*, ed. Warnock, M. New York: New American Library.
O'Neill, O. 2002. *Autonomy and trust in bioethics.* Cambridge: Cambridge University Press.
O'Neill, O. 2004. *Rationality as practical reason*, in The Oxford Handbook of Rationality, ed. Mele, A.R., and P. Rawling. Oxford: Oxford University Press.
Plato. 1984. *The Republic*, ed. and Trans. H.D.P. Lee, Middlesex: Penguin Books Ltd.
Rawls, J. 1999. *A theory of justice*, Oxford: Oxford University Press.
Savulescu, J. 1999. Desire-based and value-based normative reasons. *Bioethics* 13/5.
Scally, J. 1995. *Whose death is it anyway?* Dublin: Basement Press.
Scoccia, D. 1989–1990. Paternalism and respect for autonomy. *Ethics*, 100/2.
Slowther, A. 2007. *The concept of autonomy and its interpretation in health care, Clinical Ethics*, vol. 2, 4.
Stirrat, G., M., and R. Gill. 2005. Autonomy in medical ethics after O'Neill". *Journal of Medical Ethics*, 31.
Szasz, T.S. 1960. The ethics of birth control; or, who owns your body? *The Humanist* 20.
Szasz, T.S. 1998. *Involuntary mental hospitalization: A crime against humanity*, in Classic Works in Medical Ethics, ed. G. Pence, Boston, MA: McGraw-Hill.
Takala, T. 2007. Concepts of 'person' and 'liberty', and their implications to our fading notions of autonomy. *Journal of Medical Ethics* 33.
Tan, J.O.A. 2003. Competence to refuse treatment in anorexia nervosa. *International Journal of Law and Psychiatry* 26.
Tan, J.O.A., Hope, T. and A. Stewart. 2003. Anorexia nervosa and personal identity: The accounts of patients and their parents. *International Journal of Law and Psychiatry* 26.
Thomson, J.J. 1986. *A defence of abortion*, in Applied Ethics, ed. P. Singer. Oxford: Oxford University Press.
Verkerk, M. 1999. A care perspective on coercion and autonomy. *Bioethics* 13.
Wolff, R.P. 1969. On violence. *Journal of Philosophy* 66.
Wolff, R.P. 1970. *In defense of anarchism.* New York: Harper and Row.

Chapter 2
Enforcing Medical Treatment to Keep a Person Alive: The Problematic Case of Anorexia Nervosa

2.1 Introduction

In order to introduce the main problems present in debates over the [mis]use of enforced medical treatment, I will begin my specific analysis of controversial cases involving this issue by focusing on Anorexia Nervosa. The particularly controversial nature of the current way of dealing with Anorexia Nervosa stems from the question of whether or not we should consider anorexics autonomous enough to refuse medical treatment, given that Anorexia Nervosa is generally classified as a mental disorder. In this chapter I will more closely consider this approach, attempting to establish whether or not Anorexia Nervosa can be classified as a mental illness. Further, that being the case, I will ask to what extent this aspect can undermine the patient's competence when reaching decisions over the acceptance or refusal of naso-gastric treatment. Before moving into the philosophical sphere of the discussion, however, a more accurate examination must be carried out of how and in which ways this epidemic condition affects its sufferers.

2.2 Anorexia Nervosa: An Insight to a Contemporary Drama

In his book *Psychopolitics*, Peter Sedgwick[1] relates his dismay when, as a young, left-wing "active partisan" he discovered that from a leftist point of view issues related to mental illness were virtually non-existent, as it was the fashion

[1]Sedgwick (1982), p. 4.

M.D. Garasic, *Guantanamo and Other Cases of Enforced Medical Treatment*, SpringerBriefs in Ethics, DOI 10.1007/978-3-319-22653-8_2

of the time to deny the very fact that people do suffer from various mental disorders.[2]

Similarly, I have come to observe an inconsistency in the application of the principle of autonomy and respect for individual choice in cases of refusal of treatment in Anorexia Nervosa as well as in other mental disorders. By inconsistency, I mean the irregularity of the implementation of the principle of autonomy. This inconsistency in applying the same notion to relatively similar cases in extremely different ways is frequently evident in liberal societies such as the UK, the US and—to a certain extent—Italy, where individual choice and autonomous decision are vehemently defended under "normal circumstances". I think it is time for us to make the same mature step and understand an inconvenient truth about the processes currently at work in cases of refusal of treatment. But first we need better to understand what Anorexia Nervosa is.

Anorexia Nervosa is a specific version of those recently emergent illnesses,[3] namely Eating Disorders, that have increasingly come to affect Western and Westernised countries. In the past 30 years all kinds of Eating Disorders have seen sufficient incremental growth as to suggest a need for urgent attention to this problem. Without wanting to underplay the importance of problems such as Obesity, Binge Eating and Bulimia Nervosa,[4] this work will focus on Anorexia Nervosa alone.

Before describing the symptoms of Anorexia Nervosa in more scientific terms, it is important to underline one aspect of this condition that might easily go unnoticed and thus reduce the quality of the current analysis. By acknowledging the rise of Eating Disorders, and more specifically Anorexia Nervosa, in Western contexts, we immediately begin to prepare the ground for a linear critique of the illness not in medical terms but rather in socio-historical ones. A valuable contribution to this analysis can be achieved by reference to Daniel Callahan's False Hopes,[5] which details the undeniable truth that the Hippocratic Oath is in fact applied differently in similar cases. As Callahan explains, medicine cannot be considered to be value-free: its applications, priorities and taboos are deeply embedded in the governing power.[6]

---

[2]Certainly the Anti-psychiatric movement led by Thomas Szasz was very much in line with this idea, even though the bottom line was perhaps not to deny entirely the existence of some kind of dysfunction in the mind of certain people, the core revolution that the movement wanted to provoke was to stress the "mechanical" aspect of brain malfunction; i.e. it was curable with appropriate medicines rather than through the reassessment of the values of the individual.

[3]A definitive assessment of when Eating Disorders emerged is not the remit of this work, and for reasons of simplicity I will accept the standard date of the 17th Century as the beginning of these kind illnesses. To understand the impact of such illnesses see: Kelly et al. (2009), pp. 97–103.

[4]Not everyone agrees that Obesity is an Eating Disorder, but for a closer look at the current debate over this and other aspects of Eating Disorders see, amongst others: Fairburn and Brownell (2002); Palmer (2003), pp. 1–10.

[5]Callahan (1998).

[6]Di Paola and Garasic (2013), pp. 59–81.

Hence we should consider his critique towards Western medicine, which he finds to be too aggressive and too dependant upon the demands of a capitalist market that wants to solve its problems through the exacerbation of the conflicts of principles that it was responsible for in the first instance.[7]

Capitalism needs autonomous agents to be "free" to make their decisions, particularly regarding what to buy and consume. So too do certain medical professionals. In line with the idea that the market decides provision, we have ended up with surgeons suggesting morally dubious aesthetic operations. However, as long as we can say that the autonomous, competent citizen should be allowed to choose freely amongst the available options, little can be done to prevent or even to dilute this profit-based understanding of medicine.[8]

What creates problems with the possibility of revising such an attitude is that it constitutes a pillar of most societies that have attained a certain level of development; any such internally directed critique could spread to other areas of the same system, ultimately threatening to destabilise the very foundations of the consumerist society in which we live and in which the cases considered were able to take place.

The need for autonomy to be so prominent in bioethical contexts results from its political value. Once the role of autonomy as the leading principle in bioethics is understood—an understanding to which this work aims to contribute—we will be able to embrace a new vision of autonomy that will help us to deal with relatively similar cases in the most appropriate manner without resorting to a patently biased interpretation of this notion. For the time being, suffice it to say that, given the internal readjustment that Western society has undergone in recent times—giving rise to greater self-criticism with regard to past actions and inactions—the analysis of Anorexia Nervosa (and other Eating Disorders) began from a contested position, making it impossible from the outset to claim objectivity for any "scientific" analysis.

2.3 The Conceptualization of Anorexia Nervosa by Medicine, the Law and the Sufferers

Despite being the psychiatric illness with the highest mortality rates,[9] Anorexia Nervosa remains paradoxically the one condition that has managed to produce the least effective countermeasures to its impact. One of the main reasons for this peculiar situation lies in the crucial factor that makes Anorexia Nervosa unique: the vast majority of anorexics do not commit themselves to escape the illness. On the

[7]Whether intentionally or not, it seems that a common critique of extreme consumerism is shared by Daniel Callahan and Michel Foucault. For reason of spaces the present investigation will not develop this connection any further, but for the purposes of this work, it will suffice to underline the role that consumerism has in medicine in contemporary Western society.

[8]This is particularly true in the US, while not so evident in Europe –especially in Northern European countries such as Norway for example.

[9]Ramsay et al. (1999), pp. 147–153; Franko et al. (2004), pp. 99–103.

contrary, their embracing of the condition as a vital part of their identity results in an additional layer of ethical dilemmas that all those concerned with Anorexia Nervosa have to face. As highlighted in one study carried out by Jacinta Tan, Tony Hope and Anne Stewart: "the decision to accept treatment can become heavily loaded with the implication of giving up a part of themselves, which can affect their decision."[10]

From an historical perspective, the term Anorexia Nervosa—the most common way of referring to this condition both in English and in the international debate—was first introduced in 1873. Even though it remains unclear who first coined this term,[11] it is widely accepted that Charles Lasègue did carry out numerous studies on this Eating Disorder, defining it most commonly as "anorexie hystérique"[12] (hysteric anorexia) with all the sexist implications that such a definition entails. It is perhaps also for this reason that Mara Selvini Palazzoli would prefer the term "anoressia mentale"[13] (mental anorexia), because, on top of avoiding scientific confusion, it would also detach Anorexia Nervosa from a common inclinations to link the illness only to women. We can see quite easily that this reading is erroneous as in the last decade the percentage of males affected by Anorexia Nervosa in Western countries has increased to 8 % of the overall cases,[14] a figure which continues to rise.[15] All of the definitions listed above, however, have as their key word anorexia—etymologically meaning "lack of appetite"—which also constitutes the most common popular and media referent. However, as Simona Giordano points out, the illness does not express itself through the absence of appetite in the sufferer: the individual does have the "normal" input of feeling hungry—the presence of appetite—but she will force herself to resist it as proof of her self-discipline. She will become obsessed with food and, at the same time, with exercising her capability to resist the temptation of eating.

In this light, it should not come as a surprise to the reader that, in the vast majority of cases, the sufferer represents the prototype of a "successful individual". She would be first in class, a hard worker, striving for perfection. This "psychological identikit" is obviously limited, and it does not pretend to achieve the unachievable by defining in scientific terms the average anorexic profile. However, I believe that it is important to highlight certain common characteristics of sufferers —also in broad non-medical terms—to include a wider group of people in the analysis in which this work intends to engage.

---

[10]Tan et al. (2003), p. 546.

[11]Simona Giordano suggests in her book that this might instead have been William Gull. Giordano, Op.Cit., p.18.

[12]Lasègue et al. (1873), pp. 265–266 and 367–369.

[13]Selvini Palazzoli (1998).

[14]Fichter and Krenn (2003), p. 369–383.

[15]For example, in their report *"Treatment Decision-Making in Anorexia Nervosa"*, Jacinta Tan, Anne Stewart and Tony Hope reported an increase in the figure of male anorexics to 10 %. (p. 3) available at : http://www.psychiatricethics.org.uk/ANwebreport/report.pdf [accessed on 4 January 2015].

I am well aware that the present exploration of the clinical dimension will necessarily prove severely limited but, both for lack of space and of professional competence, this work cannot investigate the medical dimension of mental illnesses—Anorexia Nervosa more specifically—in great depth. I am confident, however, that many interesting and valuable works have been produced in recent years that allow a particularly interested reader to expand their knowledge on the topic.[16]

In the most recent version of the International Classification of Diseases (ICD) produced by the WHO, Anorexia Nervosa can be found under "*mental and behavioural disorders*" (Chap. 5), and more specifically within the section covering behavioural syndromes associated with psychological disturbances and physical factors. The definition as presented reads:

> A disorder characterized by deliberate weight loss, induced and sustained by the patient. It occurs most commonly in adolescent girls and young women, adolescent boys and young men may also be affected, as may children approaching puberty and older women up to the menopause. The disorder is associated with a specific psychopathology whereby a dread of fatness and flabbiness of body contour persists as an intrusive *overvalued* idea, and the patients impose a low weight threshold on themselves. There is usually undernutrition of varying severity with secondary endocrine and metabolic changes and disturbances of bodily function. The symptoms include restricted dietary choice, excessive exercise, induced vomiting and purgation, and use of appetite suppressants and diuretics.[17]

Even though the clinical criteria highlighted by the ICD have a measure of undeniable scientific accuracy—insomuch as is possible in medicine—it would be to offend the reader's intellectual ability not to acknowledge that in this description there are present numerous value judgements that I opt not to emphasise. I am certainly not claiming that the idea of thinness in anorexia is not *overvalued* by those individuals suffering from it, but it is unclear where we should draw the line between a noxious attitude towards life and a situation in which we can begin to speak of mental illness. For example, tobacco and alcohol abuse also figure within the "mental and behavioural disorders due to psychoactive substance use" section of Chap. 5, but there is no mention of the value that these substances are guaranteed in terms of socio-cultural acceptability and common usage. In other words, an alcoholic is presented in this description as a sane person who becomes sporadically "insane" due to the use of alcohol, or in more technical terms, he enters a phase of chronic alcoholism, but there is no direct attack on the value of alcohol itself. As a result, while on the one hand the value of thinness (strongly present in our society, tolerated and encouraged most of the time) is deemed to be "overvalued", on the other hand the same does not occur with the value of more damaging phenomena such as drunkenness and/or alcoholism. The reason behind such a discrepancy in relatively similar cases has to do with the fact that alcohol remains central to so many cultures and countries around the world that a full-scale attack on it would be too destabilising to a number of other institutional certainties that authorities do not

[16]See amongst others: Kaplan and Woodside (1987), pp. 645–653, Carney et al. (2008), pp. 199–206, Tureka et al. (2000), pp. 1806–1810.
[17]WHO (2015a). My emphasis.

want to see called into question. This statement should not come as a surprise to the reader as it has already been affirmed that the intention of this book is to reveal that such inconsistent dynamics are particularly strong in cases concerning the application of the notion of autonomy. It follows that, differently from widespread diseases such as alcoholism, Anorexia Nervosa can be expressly attacked because it affects a relatively low number of people and, most importantly, any attempt to save the lives of its sufferers does not jar with modern values. It is important to notice, however, that even accepting this reading as valid a clash would still exist. That would be the inconsistent use of terminology, serving to preserve that stability that authorities desire but that has to do with power rather than with the real nature of the illness.

In this light, two aspects of the WHO's account of Anorexia Nervosa deserve attention. The first point I want to raise is a provocative one. It is interesting to underline that, following a logic of exclusion often used in schematic and relatively scientific methods, there exist grounds to affirm that, when moving from the more general group of disorders towards the more specific one, the "mental dimension" of the disorder has been cast aside to leave the focus on the "behavioural dimension". Of course, this should be seen as a clinical categorisation of mental illnesses that aims to describe the disorder, hence behaviour—intended in the broadest sense-emerges as the main feature of Anorexia Nervosa.

However, the same logic could well prove the opposite: the definition has to focus on—and negatively emphasise—the anorexic [mis]behaviour in order to legitimise its reading of this very particular mental state as a mental disorder. Obviously, though, the fact that society does not consider a certain behaviour as rational, or even virtuous, does not function as a justification for classifying that particular state of mind as a threat to an individual's competence and autonomy. Otherwise, by parity of reasoning we should also stop drinkers and smokers from continuing in their "behavioural disturbances"! The overall perception evident in this description raises additional questions regarding the current situation which find their echo in other unconvincing contributions, leading to a more technical second point.

When reading more carefully the ICD's section on Anorexia Nervosa, there is a peculiarity not immediately evident on first reading: the definition does not apply to all cases of Anorexia Nervosa. In the very beginning of section F50 (concerning Eating Disorders) there is a list of which variants of these disorders are excluded, within which figures Anorexia NOS.[18] NOS stands for Not Otherwise Specified and is normally used for more general Eating Disorders,[19] a puzzling definition when considered alongside Anorexia. More precisely, if there are insufficient grounds to state with relative certainty that the disorder fits the definition of Anorexia Nervosa, how can it be then approximated to Anorexia NOS?

[18]WHO, International Classification of Diseases (ICD), ibid.

[19]See, amongst others: Fairburn and Harrison (2003), pp. 407–416; Eating Disorders: Anorexia, Bulimia & Eating Disorder NOS (2015).

The answer to this question comes also from the WHO's ICD schema, which in its subsequent blocks on "disorders of adult personality and behaviour", provides some material valuable to the sceptical reading developed here:

> This block includes a variety of conditions and behaviour patterns of clinical significance which tend to be persistent and appear to be the expression of the individual's characteristics. […] They represent extreme or significant *deviations* from the way in which the average individual in a given culture perceives, thinks, feels and, particularly, relates to others.[20]

It seems sufficiently clear that here the superiority of the authorities in charge—in contradiction of the [mis]judgement of the individual- is not explicitly affirmed. It follows therefore that there is a reluctance to state clearly that certain choices are not the expression of the individual, but rather that they appear to be so. This ambiguity contributes to the undermining of the respect for the clinical data analysed and supports the aim of this work in demanding a more coherent and credible way of dealing with controversial cases that revolve around the issue of autonomy.

In relation to this unconvincing use of psychiatry to justify enforced treatment—but more generally to legitimise its own authority- in her influential book *Understanding Eating Disorders*, Simona Giordano has a very interesting section in her book that examines what she calls "the Fallacy of Psychiatric' Explanations'"; this notion deserves to be considered in greater depth. Interestingly enough, the focus of her discussion is schizophrenia, the very same mental illness that will be considered in the next chapter's exploration of the Singleton case. Giordano's argument is both very simple and also very strong: in its explanation of the symptoms and effects of a mental illness psychiatry often uses an approach that fails to be logically acceptable. The logical error comes from the tautological justification given in contexts where instead the authorities involved should have the courage to accept—and publicly admit- their limits. Giordano's scheme (Fig. 2.1) allows us to understand the logical fallacy applied to Eating Disorders.

Giordano's interesting conclusion in this section of her important work, points out that, if we accept and establish that—in the vast majority of cases-[21] psychiatry can only give a descriptive picture of the mental illness, it follows that such mental disorders (including of course Anorexia Nervosa) do not compromise the autonomy of the person in question.

As she writes:

> In the majority of cases when it is said that a person has a mental illness, what is meant is that she manifests some disturbances. In most cases the psychiatric diagnosis is only a short cut to describe a pattern of disturbances: it has no explanatory value. In all cases in which the diagnosis merely has a descriptive value (and this is the majority) it is simply not true that 'mental illness' jeopardizes people's autonomy. Mental illness is a 'description of events', and as such it does not and cannot jeopardize 'autonomy'.[22]

[20]WHO, International Classification of Diseases (ICD), Op. Cit. My emphasis.

[21]Giordano (2005), pp. 68–69.

[22]Giodano, S., Op.Cit., p.70.

A Fallacy in Some Explanations of Anorexia Nervosa

Anorexia Nervosa

=

(a clinical term that) refers to/summarizes a number of disturbances
(loss of weight over stated limits, amenorrhoea, etc.)

(proper definition)

Question 1: Why have you received the diagnosis of anorexia nervosa?
(or: 'Why are you-called-anorexic'?)
Answer 1: Because you manifest the following disturbances:
loss of weight over stated limits, amenorrhoea.

(proper answer, logically correct)

Question 2: Why do you manifest the following disturbances?
(or:'Why do you diet, you have amenorrhoea...?')

Answer 2a: We are trying to understand it.

(proper and true answer)

Answer 2b: Because you suffer from anorexia nervosa.
(tautological answer)

=

You manifest the following disturbances because you manifest the following disturbances (having anorexia nervosa, in fact, means that you are manifesting the following disturbances).

Fig. 2.1 A Fallacy in Some Explanations of Anorexia Nervosa

Given the sceptical nature of this consideration, in the next section we will focus more closely on the definition of mental illness. We will then apply it to Anorexia Nervosa in order to criticise the justifications used to define Anorexia Nervosa as a mental illness, implying also a consideration of the consequences of this general consensus.

2.4 The Tension Between Competence and Mental Illness in Anorexics

Although not directly defining mental illness, the WHO constitution describes a person as in good mental health not only because of the mere absence of mental disorder. As an extension of this approach, the WHO website reads:

> Mental health is a state of well-being in which an individual realizes his or her own abilities, can cope with the normal stresses of life, can work productively and is able to

> make a contribution to his or her community. In this positive sense, mental health is the foundation for individual well-being and the effective functioning of a community.[23]

By the vocabulary used, it seems evident that the definition above points out two implicit aspects of mental illness (the absence of mental health). First, the capability to produce (a central notion in a capitalist society) is a crucial factor in establishing whether or not a person can be considered sane. Second, the actions of the individual must also be functional to the community. These variables, however, appear to be more political than medical.

Continuing with this deconstructive approach, and being provocative for the sake of the argument, one could even attempt to defend the idea that Anorexia Nervosa needs to be seen as a mental illness in order to avoid uncomfortable situations of biased judgements over relatively similar cases. After all, as Giordano rightly points out:

> The person with an eating disorder is far removed from the common idea of the 'insane' and may be a skilled and competent person in virtually all areas of her life. [...] If people are normally entitled to choose their lifestyle, however dangerous or irrational it may appear to others, why should not people be able to choose what and how they want to eat?[24]

Even more so, the fact of having Anorexia Nervosa outside of the standard ways of classifying a mental illness can surely be argued to be convenient for a certain project. In fact, having Anorexia Nervosa as non-classifiable "normal" case of refusal of treatment could be seen as a very useful way out for the judicial system in situations where the role of mental illness, competence and autonomy can be used in inconsistent ways to favour the prevailing political trends.

Without wanting to enter into a deep technical debate on the definition of mental illness, I will now aim at highlighting the main implications for decisional processes of suffering from such an illness. In other words, I will take Anorexia Nervosa to be a mental illness, but I will question the meaning of precisely this definition.

The key aspect of this consideration will be to establish whether or not a mentally ill patient can still be deemed competent. In this respect we should consider the view of Thomas Szasz,[25] amongst others. According to this view, if we were to consider mental illness an actual illness, it would be one of the brain, not of the mind. With such an approach towards cases of Anorexia Nervosa, for instance, it should become clear that once the incapability of the doctors to improve the situation, or more simply to cure the illness, is accepted, the decisional power should return to the competent patient. This idea will be analysed in greater depth in the next section of this chapter, but, before taking that path, we shall consider an additional aspect relevant to a full understanding of Anorexia Nervosa. In relation to the evaluation of Anorexia Nervosa as a pathology particularly linked to female characteristics, we must understand these as gender-specific limits shaped by historical injustices. Helen

---

[23]WHO (2015b).

[24]Giordano, S., Op.Cit., pp. 30–31.

[25]Szasz (1972).

Malson's very interesting work, *The Thin Woman*, provides an analysis of the "genealogy of anorexia", pointing out that, despite recent improvements in the relationship between genders that have given more respect to women, there is still an acceptance of the intrinsically masculine concept of "healthiness".[26]

Such an acknowledgement is certainly worthy of attention, but, despite supporting Malson's application of Foucault to the current analysis of Anorexia Nervosa—and the resultant belief that to understand it fully we cannot limit ourselves only to the result of a historic-medical discourse- I believe that certain characteristics of Anorexia Nervosa are objective realities that signify illness regardless of their links to a specific gender. As proof of its "intergenderness" it would be worthwhile to consider once again that in recent years the number of males affected by Anorexia Nervosa has drastically increased and can sometimes even produce more problems related to the specific biological structure of male sufferers.[27]

This recognition leaves us with two considerations to take into account: the first is that, if we had to accept the conservative male-centred view of Anorexia Nervosa, this would be perhaps a good occasion to understand that if the illness is "transmittable" between genders, the problem lies in the external factors that produce the precondition for Anorexia Nervosa to develop (obsession with body image, need to prove one's will power). This accepted, the conservative view would be knocked off its chauvinist pedestal. The second consideration that deserves attention is that, as for Szasz, the mere awareness of the fact that something was abused in the course of history in order to prolong the continuation of an injustice is not sufficient reason to refute the scientific validity of those data that we currently have. As a matter of fact, Szasz himself did not claim that psychiatry does not exist, but only that we should reshape its use.[28]

We have already explained in the previous chapter the definition of competence vis-à-vis the notion of autonomy and its legal and medical status. Here, we will look at this definition in closer relation to Anorexia Nervosa. To evaluate the impact of Anorexia Nervosa on the competence of those refusing naso-gastric treatment, it has been accepted that Anorexia Nervosa is a mental illness. In arriving at this acceptance, however, the question that we have raised focused on affirming that even given such a scenario there is no clear evidence that the incapability to judge competently in decisions related to food would necessarily jeopardise the competence of the anorexic in any given context.

As we have seen above, it is not entirely clear whether or not the anorexic sufferer can be claimed to be incompetent in every context. In truth, it appears well accepted that they are indeed competent in most cases. They are perhaps incompetent when it comes to food, but not when asked about their quality of life. This is

[26] Malson (1998), pp. 47–48.

[27] In this respect it is interesting to note the different reaction that females and males have towards involuntary treatment. A good example of this distinction is Silber et al. (2004), pp. 415–418.

[28] Szasz, T., The Myth of Mental Illness, Op. Cit.

clearly the main problem to deal with: if they are competent, can we still override their will and force-feed them?

Some positions would argue that there are cases, even if very small in number, where such refusals should be heeded, and the reason for such an affirmation is that in these given instances the patients would be in a position to make a competent decision. One very common position would then argue that patients suffering from Anorexia Nervosa are not capable of making any competent decisions regarding feeding or, more generally, any issue relating to food. These views accept this position but highlight that in cases of naso-gastric treatment concerning "experienced" and relatively "stable" patients (persons that have already been through such therapy and that are in no immediate danger of death) the issue to consider relates not to food but rather to concerns over quality of life. These patients would be able to make competent decisions, because these decisions would not be related in any direct way to food. Critiques of this point are based on the further development that affirms that if we recognise anorexics as competent, we should be ready to affirm their autonomy as well. Such critiques and their counterarguments shall be addressed through the analysis of the key concepts of competence and autonomy. As Heather Draper suggests in her paper:

> What needs to be established, and what is very difficult to establish in the case of anorexia nervosa, is whether the person with anorexia nervosa is an autonomous agent who is incompetent to make some judgements, or a non-autonomous agent who is competent to make some judgements.[29]

Yet, it is important to take into account another crucial factor: in the cases considered, doctors are not expecting the situation to improve, their intention is only to postpone death insomuch as possible. Under these conditions, however, it seems obvious that the moving principle behind the decision not to interrupt a treatment or switch off a vital machine has to do with the moral view of the doctor on the matter. But should it be so? Should the will of the patient not be respected if the actual consequences of the most extreme decision would only result in the acceleration of a process otherwise incredibly painful? After all, the Anglo-American norm in medical contexts it is to accept the decision of the competent patient as decisive, including when their decision would result in death.

2.5 Are We to Enforce Medical Treatment in Cases of Anorexia Nervosa?

The recent 2012 ruling in the UK where it was affirmed that an anorexic woman should have been force-fed against her will[30] has revived the debate over the permissibility of such procedure in cases of Anorexia Nervosa. As a moral

[29] Draper (2003), p. 4.

[30] Dyer (2012), p. e4232.

justification for refraining from artificially feeding a patient suffering from Anorexia Nervosa, it could be argued that it would be a practical form of the doctrine of double effect. In fact, this interesting argument is pursued by Fiona Randall and Robin Downie in their book *Palliative Care Ethics*:

> the doctrine of double effect which relies on a moral distinction between intended and foreseen events allows the use of measures to relieve suffering even though they carry a significant risk of shortening life.[31]

I shall argue that adapting this approach to the interruption of naso-gastric treatment would produce the effect of defending this option as functional to the reduction of the patient's suffering with the unintended result of letting the self-same patient die. For the sake of the argument, it might be claimed that from a utilitarian point of view it could even be justifiable to force treatment on anorexic patients because their internal suffering would still produce less "moral" damage to the consciences of the persons around them (family, friends, and doctors) than would their death. This approach, however, would deny the centrality of ensuring that the patient's autonomous decisional power be defended in all cases where the patient's competence has been established.

In fact, I want to suggest that one of the justifications for the disparity of strictness in accepting the will of the patient as morally permissible and based on competence may well be linked to the possible consequences of denying such freedom to translate choices into actions. The reason is self-evident: while in the case of terminally ill patients the hope for recovery has completely disappeared and nothing will prevent the patient from dying, in the case of Anorexia Nervosa the hope may always exist, including for the patient herself. To not accept any refusal is often seen as a way of gaining time in which the patient might "come to her senses" and move away from a condition of extreme Anorexia Nervosa towards a less extreme stage of the illness at least. However, it is through the acceptance of such a strategy (that may often be rooted in noble intentions) that I hope to have highlighted what does not satisfy me about the present discrepancy between different types of treatment refusal, all of which would eventually result in death. For the situation just considered would imply a level of paternalism on the part of the doctors that we claim to be unjustified when the patient has the capability to make a competent decision. In other words, if the patient is found to be competent, we must allow her to pursue her destiny despite our concerns over the "chances of success" were any refusal of treatment to be accepted. We should be ready, as Giordano says, to make the "brave claim".[32] Admittedly, this is not a decision to be made light-heartedly and for this reason in the next section of this chapter we will shift our attention on to the unique complexity of the problems surrounding Anorexia Nervosa.

---

[31]Randall (1999), p. 127.

[32]Giordano, S., Op.Cit., pp. 246–250.

2.6 The Biopolitical Reasoning for Keeping Anorexics Alive

The deep-rooted dilemma in Anorexia Nervosa is that it is a very peculiar condition which, in developing as early as the age of twelve (this figure falls each year as the pressure on youngsters grows), makes it extremely difficult to ascertain precisely when the patient has recovered from the mental illness, because in most cases the mental illness itself has evolved as part of their own personality and way of being. We could say that in some ways abruptly breaking this link with a part of their selves could prove seriously destabilising, a point that should probably be given greater consideration than is usually the case. A metaphorical representation of what it means to develop Anorexia Nervosa might be the science-fiction-type situation where some children grow up with tinted glasses fixed to their eyes. In time, their particular way of seeing the world (through green-tinted lenses, say) will become their only accessible and conceivable reality. With this simple yet hopefully valuable scheme in mind, two considerations arise in relation to Anorexia Nervosa.

The first consideration concerns the potential damage done by removing the sunglasses too abruptly from the eyes of the patient. As the reader might know from personal experience, such an action is always followed by a moment of temporary blindness. In the imaginary scenario portrayed above, the situation entails an exponential increase both of the time of exposure to the sun and the time in which the eyes adapt to seeing the world through green filters. As a result, it should be easily understandable that a precipitous choice—aiming to show the true colours of the world to the patient through sudden removal of the "anorexic sunglasses"-might result in a more damage than benefit, at least in the short term.

The second aspect to consider—and the one more closely linked to the purposes of this work—relates to the value that we assign to the role of the green-filtered sunglasses when establishing the level of competence of the individual in question. More specifically, crucial is the certainty with which we can affirm that this distortion of reality impairs the person's ability to analyse competently important features other than colour; to deny respect for general competence on the grounds of possible incompetence in a certain domain would hardly be justifiable. Continuing with our metaphor, then, we could say that, on the one hand, it would be reasonable to accept that in the condition described it would be unrealistic to expect the person with sunglasses to be able to distinguish between two objectively distinct shirts (one green and one white) that to her green-filtered eyes will result undistinguishable.

On the other hand, however, would it not be unreasonable to claim instead that due to her sunglasses, if put in the condition of having to do so, the person in question would not at least try to dodge a (grey) stone thrown at her? Instinctively, no one would deny that the absence of competence in regard to the (partial) colour-blindness of the person would not still represent a sufficient impediment substantially undermining the self-preservative nature of the individual who will do anything in her power to avoid the potential pain caused by the stone. Though simplistic, this example could well function as a launch pad to enter into a more sophisticated discussion of this peculiar—and controversial—aspect of Anorexia Nervosa: the shaky ground on which rests the assessment of partial incompetence.

To avoid a serious confrontation on this topic, with all its potential consequences in the biopolitical sphere, many proposals have been touted. For example, the possibility of using nocturnal naso-gastric treatment[33] is significant and worthy of particular attention as it attempts to reduce the clash between the medical obligation to treatment and the explicit overriding of the patient's will. But while it might succeed in making this contrast less violent, it still fails to provide a satisfactory solution.

We might feel entitled to feed the patient while she is asleep without asking her permission, perhaps without even informing her of the treatment in order to avoid problems related to standard naso-gastric treatment. To do so, however, would entail the sidelining of the question of whether or not the patient is competent or not, the patient being left bereft of any possibility to decide how to deal with her situation. As such—aside from the purely technical aspects—the nocturnal naso-gastric treatment does not differ in any significant way from a standard paternalistic approach that would naturally presume the incompetence of the anorexic patient.[34]

The brief analysis produced in this chapter will lead us back to the initial question that haunts those attempting to find an acceptable solution to the ethico-legal problems associated with Anorexia Nervosa: what should authorities do when faced with such cases? How should we, as a society, behave in such an ambiguous situation? Where to draw the line of respect for freedom of choice and for life?

In her article 'Anorexia: a Role for Law in Therapy?'[35] Terry Carney focused on a very practical way of dealing with the issue, namely ensuring that law would guarantee the preservation of life in somuch as possible. She writes:

> It (is) hard to reject a role for law in the authorisation of the use of coercion in some form in the case of emergency or life-saving interventions for severe anorexia nervosa. But […] it is equally difficult not to accept that a guardianship-type order/jurisdiction[36] has a legitimate role as well, and indeed should serve as the preferred initial measure when legal intervention is required.[37]

Thus, despite having an intuitive leaning towards the preservation of life as the ultimate duty, the law should first respect its own limits, and accept that at this stage in the majority of Anglo-American legal systems the principle of autonomy resists any attempt to be diluted.

---

[33]See footnote 27 above and Robb et al. (2002), pp. 1347–1353.

[34]Halse et al. (2005), pp. 264–272.

[35]Carney (2009), pp. 41–59.

[36]Here Carney refers to a specific term used in Australian contexts in relation to a third person (a guardian/tutor) deciding on behalf of the patient in question. Obviously, if the anorexic is found to be competent enough to make a decision, the guardianship remains with her.

[37]Carney, T., Op.Cit., pp. 41–59.

This awareness, combined with the commonsense intuition that it would be morally wrong to allow the loss of life of certain anorexics (many of whom will later prove grateful for having received enforced treatment) for the sake of respecting this self-imposed predominance of autonomy, should lead us to ask if this system based on an individualistic version of autonomy is indeed as suitable as we currently believe it to be.

2.7 Conclusion

In the course of this second chapter we have moved the analysis of the notion of autonomy in bioethical cases from a more theoretical discussion towards a more empirical, fact-based approach. More specifically, our focus has turned to controversial cases of enforced naso-gastric treatment in Anorexia Nervosa, developing further—and in greater contextual depth—the concepts of competence, autonomy and mental illness relevant to all of the cases considered in this work. Through an investigation that has brought to the fore the medical peculiarity of Eating Disorders —and more specifically Anorexia Nervosa—when evaluated in terms of autonomy and competence, it has been pointed out that patients suffering from Anorexia Nervosa cannot be so easily separated from their illness as can those undergoing most other medical procedures. As shown with the arguments sustained and convincingly articulated by Giordano, we have shifted the debate over the legitimacy of enforced treatment in Anorexia Nervosa into a field that questioned more vigorously the limits that authority can (or should) have in relation to the values of individual. In doing so, we have reinforced the conviction that, while Anorexia Nervosa might not jeopardise the level of the patient's competence to such an extent that enforced treatment can take place under current legal and moral standards, the unacceptability of the refusal of treatment in Anorexia Nervosa is related to the impact that such an acknowledgement would have on wider societal values. By referring extensively to previous researches and perspectives, it was not the intention of this chapter to be particularly original in its content. Rather, its function was to introduce the reader to the multilayered problems related to enforced medical treatment through an in-depth analysis of cases of Anorexia Nervosa—as they represent a unique example of the tension between respect for patient's autonomy (especially if assessed to be competent in all but one field), medical concerns and political choices. The way in which these three aspects interact in different cases is the central theme of the book and Anorexia Nervosa represented the best way to highlight the limits of the conflicts that patients, doctors and political authorities need to face when dealing with any kind of enforced medical treatment. The inconsistent use of autonomy as a function of its political context will be further analysed in the next chapter.

References

Callahan, D. 1998. *False Hopes: why America's quest for perfect health is a recipe for failure.* New York: Simon and Schuster.

Carney, T. 2009. Anorexia: a role for law in therapy? *Psychiatry, Psychology and Law* 16(1): 141–159.

Carney, T., Tait, D., Richardson, A., and S. Touyz. 2008. Why (and When) clinicians compel treatment of anorexia nervosa patients. *European Eating Disorders Review*, 16: 199–206.

Di Paola, M., and M.D. Garasic. 2013. The dark side of sustainability: On avoiding, engineering, and shortening human lives in the Anthropocene. *Rivista di Studi sulla Sostenibilità* 3(2): 59–81.

Draper, H. 2003. Anorexia nervosa and refusal of Naso-Gastric treatment: A reply to Simona Giordano. *Bioethics* 17: 261–278.

Dyer, C. 2012. Severely anorexic woman should be force fed, judge rules. *British Medical Journal* 344: e4232.

Eating Disorders: Anorexia, Bulimia & Eating Disorder NOS. 2015. http://cpancf.com/eatingdisorders.asp. Accessed 4 Jan 2015.

Fairburn, C.G., and K.D. Brownell (eds). 2002. *Eating disorders and obesity: A comprehensive handbook*, 2nd edn. New York: Guilford Press.

Fairburn, C.G., and P.J. Harrison. 2003. Eating disorders. *Lancet*, 361: 407–416.

Fichter, M., and H. Krenn. 2003. *Eating Disorders in males*. In *handbook of eating disorders*, eds. Treasure, J., Schmidt, U., and E. Van Furth. New York: John Wiley and Sons.

Franko, D.L., Keel, P.K., Dorer D.J., et al. 2004. What predicts suicide attempts in women with eating disorders? *Psychological Medicine* 34: 843–853.

Giordano, S. 2005. *Understanding eating disorders*. Oxford: Oxford University Press.

Halse, C., Boughtwood, D., Clarke, S., Honey, A., Kohn, M., and S. Madden. 2005. Illuminating multiple perspectives: Meanings of nasogastric feeding in anorexia nervosa. *European Eating Disorders Review* 13: 264–272.

Kaplan, A.S., and D.B. Woodside. 1987. Biological aspects of anorexia nervosa and bulimia nervosa. *Journal of Consulting and Clinical Psychology* 55(5): 645–653.

Kelly, L., Klump, K.L., Bulik, C.M., Kaye, W.H., Treasure, J., and E. Tyson. 2009. Eating Disorders Are Serious Mental Illnesses. *International Journal of Eating Disorders* 42(2): 97–103.

Lasègue, C.E. 1873. *On histerical anorexia*. Medical Times and Gazette, 2.

Malson, H. 1998. *The thin woman: Feminism, post-structuralism, and the social psychology of anorexia nervosa*. London: Routledge.

Palmer, R.L. 2003. Concepts of eating disorders. In *handbook of eating disorders*, eds. Treasure, J., Schmidt, U., and E. Van Furth. New york: John Wiley and Sons.

Ramsay, R., Ward, A., Treasure, J., and G.F. Russell. 1999. Compulsory treatment in anorexia nervosa: short-term benefits and long-term mortality. *British Journal of Psychiatry* 175: 147–153.

Randall, F., and R.S. Downie. 1999. *Palliative care ethics*, 2nd edn. Oxford: Oxford University Press.

Robb, A.S., Silber, T., Orell-Valente, J.K., Valadez-Meltzer, A., Ellis, N., Dadson, M.J., and I. Chatoor. 2002. Supplemental nocturnal nasogastric refeeding for better short-term outcome in hospitalized adolescent girls with anorexia nervosa. *American Journal of Psychiatry* 159(8): 1347–1353.

Sedgwick, P. 1982. *Psychopolitics*. London: Pluto Press.

Selvini Palazzoli, M. 1998. L'anoressia mentale: Dalla terapia individuale alla terapia familiare, 9th edn. Milan: Feltrinelli.

Silber, T., Robb, A.S., Orell-Valente, J.K., Ellis, N., Valadez-Meltzer, A., and M.J. Dadson. 2004. Case report nocturnal nasogastric refeeding for hospitalized Adolescent Boys with anorexia nervosa. *Journal of Developmental & Behavioral Pediatrics* 25(6): 415–418.

Szasz, T.S. 1972. *The Myth of mental illness: Foundations of a theory of personal conduct.* London: Paladin.

Tan, J.O.A., Hope, T. and A. Stewart. 2003. Anorexia nervosa and personal identity: The accounts of patients and their parents, *International Journal of Law and Psychiatry* 26: 533–548.

Tureka, L., Wayne, W., Bowers, A. and A. Andersen. 2000. Involuntary Treatment of Eating Disorders. *American Journal of Psychiatry* 157: 11.

WHO. 2015a. International Classification of Diseases (ICD). http://www.who.int/classifications/icd/en/. Accessed 4 Jan 2015.

WHO. 2015b. http://www.who.int/mediacentre/factsheets/fs220/en/. Accessed 4 January 2015.

Chapter 3
Enforcing Medical Treatment to Kill: The Case of Charles Laverne Singleton

3.1 Introduction

In the course of this chapter I will highlight a very controversial way of implementing the notion of autonomy as previously considered. In October 2003[1] the Supreme Court of the United States allowed Arkansas officials to force Charles Laverne Singleton, a schizophrenic prisoner convicted of murder, to take drugs that would render him sane enough to be executed. On January 6 2004[2] he was killed by lethal injection, raising many ethical questions. By reference to the Singleton case, this chapter will analyse in both moral and legal terms the controversial justifications of the enforced medical treatment of death-row inmates. I will begin by providing a description of the Singleton case, before highlighting the *prima facie* reasons for which this case is problematic and merits attention. Next, I will consider the justification of punishment in Western society and, in that context, the evolution of the notion of insanity in the assessment of criminal responsibility during the past two centuries, both in the US and the UK. In doing so, I will take into account the moral justification used to enforce treatment, looking at the conflict between the prisoner's right to treatment and his right to refuse medication where not justified by reasonably foreseeable positive outcomes for the individual. Finally, in contrast with some retributivist arguments in favour of enforced treatment to enable execution, I will propose a possible alternative, necessary if we are to consistently uphold the notion of autonomy. It will be argued that, within the current Anglo-American legal framework, in cases of capital punishment where the inmate

This chapter has been published as a self-standing article. Garasic, M.D. 2013. The singleton case: enforcing medical treatment to put a person to death. *Medicine, Health Care and Philosophy* 16(4): 795–806.

[1]*Singleton v Arkansas*, 124 S.Ct. 74 (2003) (Cert. Denied).

[2]Available at: http://www.clarkprosecuter.org/html/death/US/singleton887.htm [accessed on 4 January 2015].

M.D. Garasic, *Guantanamo and Other Cases of Enforced Medical Treatment*, SpringerBriefs in Ethics, DOI 10.1007/978-3-319-22653-8_3

was competent at the moment of sentencing, the death penalty should be carried out as normal, since the immutability of the sentence makes it impossible to justify enforced treatment in either legal or moral terms.

3.2 The Singleton Case

In 1979 Charles Laverne Singleton killed a grocery clerk in Arkansas and was sentenced to death that same year. Once on death row, he began taking psychotropic medications to alleviate anxiety and depression. However, in 1987, his mental health deteriorated further to the extent that he claimed that his victim was still alive and that he himself was possessed by demons. Singleton was diagnosed as schizophrenic and prescribed antipsychotic medication. During the following years he oscillated between agreeing and refusing to take the medication. As a result, when he spontaneously refused to take it, it was forced on him. When he went off the medication, the paranoid and delusional behaviours returned. By 1997 antipsychotic medication had become so necessary that the prison placed Singleton under an involuntary drugging regime, subject to annual review. Under this regime, Singleton's mental health improved to the extent that the State of Arkansas authorities considered him eligible for execution, scheduling it for March 2000. Singleton then filed a petition for *habeas corpus*,[3] contending that he was only competent because of the medication he was being forced to take and that it was unconstitutional to use enforced medication to raise his competence such as to become eligible for execution. The Eighth Circuit Court of Appeals was called upon to decide whether the state could execute someone forcibly medicated in order to meet the competence requirements for proceeding with the execution. The following sections will focus on the arguments arising out of this decision.

3.3 Prima Facie Problems

The Singleton case has produced a paradoxical position on the part of the relevant authorities, since they claimed that the best outcome was for Singleton to be forced to take the medication and then executed, rather than living in psychosis and imprisonment. In order to defend the fairness and righteousness of such an interpretation, various—sometimes contrasting—principles have been invoked as proofs of its legitimacy. In order to satisfactorily take into account the multiple principles, values and laws involved in the judging process, I will divide the multilayer

[3]A legal action of English origin which has been a historically important instrument to ensure protection of individual freedom against arbitrary state action, and which can be used to seek relief from unlawful detention.

structure of the Singleton case into smaller pieces, separating the chapter into sections and sub-sections that will each focus on a relevant aspect of the justification of the final decision reached by the Eighth Circuit Court of Appeals.

Hence, to understand more appropriately the ethical dilemmas involved in the Singleton case, we shall start by looking at its *prima facie* problems. Three questions arise from the controversial position just described. First, how, if at all, can we evaluate the attainment of a satisfactory level of competence that would allow for execution? It should be noted, that Singleton's lawyer claimed his client's restored competence to be based on an "artificial sanity"[4] not related to the original individual. This aspect is important because, as will be highlighted later in the chapter, sanity is a crucial factor in the assessment of the legal responsibility of an agent in perpetrating a criminal action. This also suggests an additional problem regarding the non-continuity of the agent over time and calls into question the level of responsibility that the present agent can have for the actions of the past agent. As a result, the presence of an alternative sanity unrelated to the original agent would make the whole process of re-establishing mental competence pointless. Second, why should the state insist in curing a prisoner against his will if such an imposition would inevitably result in death? Is there a need to provide an exemplary punishment for those outside prison, or is the main aim to ensure the fully conscious suffering of the competent prisoner as an integral part of the punishment? If the latter, these reasons need to be made explicit, rather than claiming that enforced treatment is in the best interest of the prisoner, as happened in this case. Third, if the penalty cannot be changed, would it not be more logical, and perhaps more humane, to execute the prisoner no matter what his mental state is at the time of the execution, instead of prolonging his agony? After all, even when legislation allows ethics committees to override patient's informed refusal—as in Israel, for example—three conditions must be satisfied:

(1) Physicians must make every effort to ensure that the patient understands the risks of non-treatment.
(2) The treatment which physicians propose must offer a realistic chance of significant improvement.
(3) There are reasonable expectations that the patient will consent retroactively.

Of these three points listed by Michael Gross,[5] at least two of them seem not to be satisfied in the Singleton case, and it would be difficult to claim that there would be an improvement for the agent suffering the enforced treatment, and, most of all, that the patient would consent to the treatment retroactively, as he would be dead. This aspect is indeed of primary significance and will be kept in consideration throughout the chapter.

[4]*Singleton v Norris*, 319 F.3d 1018 (8th Cir. 2003) (Habeas-Competency). Available at: http://www.cognitiveliberty.org/dll/singleton_8circ2.htm [accessed on 4 January 2015, p. 16].

[5]Gross (2005), pp. 29–34.

In line with this view, in the State v Taylor[6] case there was a claim for the unconstitutionality of enforced treatment. One of the main reasons for this claim was the fact that the condition resulting from the enforced treatment was actually more damaging than beneficial to the agent. It should also be noted that the involvement of doctors and psychiatrists in such a specific case would clash with the AMA Code of Medical Ethics.[7] This aspect should be considered by the competent authorities as an important factor in evaluating the moral acceptability of the enforcement of this law. The result would perhaps undermine any positive perception of such a procedure, as the whole process represents a problematic situation for the doctors and psychiatrists implicated in the practice.[8] Although this is an important aspect to acknowledge, for reasons of space it will not be discussed any further here. Rather, I will now direct my attention to the idea that enforced medical treatment is justified on the grounds that the rights of the prisoner can be overridden in his best interest as the better of two evils. However, when the death penalty is involved, this approach becomes quite paradoxical, and if the ultimate aim of society is to execute the agent, why should we override his or her wishes if the positive outcome of re-establishing an "acceptable" mental state would abruptly disappear with death? I will aim to answer these questions in the following sections.

3.4 Neuroscience, Enforced Treatments and Other Perspectives

Before entering the core of my investigation, some clarification regarding the scope of this work should be given. A variety of issues could be considered, but not being able to discuss all of them in proper depth here, I can only acknowledge and elucidate my engagement with some of these more peripheral concerns. First of all, the Singleton case deals with an extreme form of punishment, namely, the death penalty. I am well aware that capital punishment is a debated issue in itself and that is why this chapter will be limited to an analysis of this particular case within its legal boundaries without questioning the moral justifications for their existence. The focus will therefore be directed towards the idea of restoring an agent's competence in order to punish in accordance with the degree of responsibility assigned to him or her. Obviously, the kind of punishment involved in the Singleton case makes a difference to this evaluation. However, at this stage, what should be considered as the central question to ask is: in a case of life imprisonment, rather than execution, would it be tolerable to enforce medical treatment in order to ensure the appropriate level of competence throughout the experience of the punishment? I

---

[6] *State v Taylor*, S83428.

[7] *Code of Medical Ethics*, American Medical Association, Opinion 2.06, Capital Punishment.

[8] See, amongst others: Peloso and Bandini (2007), pp. 245–266.

will aim to show that while in this latter case enforced treatment could be justified, when capital punishment is involved the evaluation changes significantly.

A second aspect that could be considered in this context is the aforementioned doctrine of double effect. Would it make any difference if the state argued that the reason for enforced treatment was mainly the health of the prisoner, and the fact that restoration of his mental capacity would result in his execution simply an unintended consequence? I think that this question should be answered negatively because during a state of "temporary" competence Singleton refused treatment. Under normal circumstances, his decision should have been respected as long as he was found to be competent, but in Singleton's case treatment was enforced in order to ensure that punishment could be carried out. Surely this aspect could be further analysed, but here I only want to show why claiming to enforce treatment on Singleton primarily for his health, rather than to allow for punishment to take place, is currently unjustified. In relation to this aspect, a third point should be taken into account. Some will argue that involuntary treatment should never take place. Again, the evaluation of the validity of such a position in absolute terms, as well as within the specific legal system considered, could be the object of a separate study in itself. In relation to the present analysis, though, I will limit myself to pointing out that a full-scale defence of the role of voluntariness in the acceptance of treatment does not differ greatly from the position of this chapter. For the ultimate intention to not medicate and then execute is in line with the acceptance of a retributivist approach to law but, at the same time, does not pretend hypocritically to affirm that we should override the prisoner's refusal of treatment in order to accommodate the need for our law to make the criminal pay.

A final, but very important, issue to consider—and even though this chapter focuses upon the role of medical inputs through the use of psychotropic medicines —is that other biomedical means[9] are currently available to restore competence in a patient in order to ensure the much needed "moral enhancement" vital to the justification of the whole procedure. Indeed, they are even considered to be morally acceptable by positions such as the one defended by Thomas Douglas in his article "Moral Enhancement".[10] The enforced restoration of competence does not only apply to extreme cases of capital punishment, but can, and often does, also include situations in which involuntary medication is used to make the defendant Competent to Stand Trial (CST).[11] Obviously, the assumption for such enforced treatment is that, under normal circumstances, an adult human being is a responsible being, and therefore liable under the law. It is the duty of authorities, therefore, to re-establish that lost "normality" in the individual in order to ensure the prompt return to standard procedures of responsibility assessment. This, at least, is the background justification for the vast majority of such treatments. But one wonders

[9]Amongst other methods such as brain surgery, TMS and deep brain stimulation, one of the most debated approaches to restoring competence is represented by Electro-Convulsive Therapy (ECT). Concerning this treatment see, amongst others: Ladds (1995), pp. 183–187.

[10]Douglas (2008), pp. 228–245.

[11]See, amongst others: Gerbasi and Scott (2004), pp. 83–90.

whether the "benevolent" capacity of these treatments to restore competence, a significant part of what constitutes a given individual, might instead function only as a way of ensuring the suffering of an "alternative agent"[12] not clearly responsible for past actions. If this is the case, would their use be still morally sound?

Again, I cannot discuss this aspect in depth,[13] but for the present investigation, it is important to note once more that we can consider the possibility of enforcing treatment in order to ensure the appropriate punishment of the responsible agent only by accepting the continuity of agency in time. Were we to call that premise into question, the whole idea of punishment would have to be revised. This component is indeed very important, for it is normally given that the continuity of the agent remains intact in time. Otherwise, it would be very difficult—if not impossible—to find a consistent way to assign responsibility for an action to a given individual. This key aspect will be analysed in the next section within the framework of the idea of punishment, albeit only briefly.

3.5 Punishment, Insanity and Responsibility

The Singleton case is particularly controversial because it managed to combine aspects of the moral and legal spheres that were already difficult to deal with in themselves. In order to disentangle this twine of background notions involved in the judgement procedure, in the following three sub-sections, I will focus separately on the notions of punishment, insanity and responsibility. To do so, the historical idea of punishment, and the correlated role of insanity in the assessment of culpability, will be considered in the first and second sub-sections respectively. In the third sub-section, I will instead take a closer look at the retributivist argument that defends the court's decision as a synbook of the two previous sub-sections, guaranteeing in this way the fairness of the judgement.[14]

---

[12]See footnote 4 above.

[13]To be more specific on the issue of the temporal dimension of responsibility, it should be noted that there is wide agreement that responsibility can be looked at in at least two different temporal directions: *backward* and *forward*. For reasons of space, I cannot discuss this point in greater depth here, but what is important to note for the purposes of this work is the general acceptance that if certain variables change over time, the assessment of responsibility can be influenced in accordance. For a more complete account of this issue see, amongst others: Vincent (2009), pp. 39–51; Kutz (2004), pp. 548–587; Duff (1998), pp. 290–294.

[14]I am aware that other ways of justifying enforced treatment could be considered. Amongst these, certainly the Hegelian idea of reconciliating Singleton to society (and to an extent to his true self) could be seen as a powerful argument. However, without denying its validity, an important clarification must be made. This thesis will not focus on these parallel approaches for two reasons: first of all, properly explaining those arguments would require a much more detailed investigation than the scope of this work permits. Secondly -as will become apparent later on in the chapter- my intention is to model this part of the work on a pre-existing article that places retributivism at its centre.

3.5.1 The Idea of Punishment

> In everyday contexts, when lawyers, judges, parents, and others are concerned with issues of responsibility, they know, or they think they know, what in general the conditions of responsibility are. [...] Is this person mature enough, or informed enough, or sane enough to be responsible? Was he or she acting under posthypnotic suggestion or under the influence of a mind-impairing drug? It is assumed, in these contexts, that normal, fully developed adult human beings are responsible beings.[15]

In the light of these remarks of Susan Wolf on the background notion of responsibility under normal circumstances, we shall now look at the idea of punishment in the Western tradition in order to contextualise better the position of Singleton before the law. There are, of course, views in philosophy that would disagree with the premises of this work. However, it is my intention,—in some ways comparable to those of Wolf and others-[16] to approach the specific case with a method of investigation that will go beyond the debate between determinism, libertarianism and compatibilism. I want to focus in very practical terms on the justifications behind this approach in law that aims to see the prisoner's competence restored before continuing with the procedure of capital punishment. A prior acceptance of the Anglo-American legal apparatus based on the use of punishment should therefore be admitted, and three considerations about the conception of punishment should be made. Firstly, if on the one hand, the agent is considered not to be responsible for his actions as a deterministic approach might suggest, there would be no reason to re-establish his competence in order to justify his execution. If, on the other hand, the agent is considered to be a free willing individual—and our current legal system indeed presumes a "normal" person to be so—then the competence of the agent becomes undeniably relevant to any assessment of the level of intentionality, and, consequently, the degree of guilt. This is what makes the distinction between *mens rea* ("guilty mind") and *actus reus* ("guilty action") crucial in establishing the appropriate punishment in a sentence.

Secondly, the continuity of the agent in time should be taken into account. As mentioned above, the connection between the time of the crime, the time of the sentence and the time of execution is what makes the Singleton case so unique and indeed so controversial. I am aware that the degree to which we can affirm continuity in the actions of an individual can be, and has been, debated. Sometimes—for example in cases of dementia-[17] we might even question the extent to which a person is the same as before, even if he remains in the same body. In a similar approach, in *Reasons and Persons*, Derek Parfit[18] criticises the dogmatic idea that the same individual at different moments in time is the representation of the same

---

[15]Wolf (2003), p. 372.

[16]See, amongst others: Wolf, S., ibid., pp. 372–387, and Bok, H., Op.Cit., pp. 130–167 both in Watson, G., (ed), Free Will, OUP, 2003.

[17]Hope (1994), pp. 131–143.

[18]Parfit (1984).

person, suggesting that a prisoner should have the punishment reduced proportionally according to the looseness of the connection with the "past self" that committed the crime. According to this view there are, in a sense, two different accountable individuals. However, without wanting to deny that such an approach can be successfully used as a theoretical basis in a rehabilitation process of wrongdoers, I shall claim that an argument in favour of non-continuity in the evaluation of the person would prevent society from producing an accurate account of an individual's actions, and make any decision perpetually changeable in accordance with mental changes that occurred in the patient. Obviously, this perpetual instability would be of little help from a legal point of view and, as a result, for the purposes of this work.

Thirdly, Herbert Hart's[19] description of the two conditions required to justify punishment must be highlighted. The first regards conviction by court: the criminal act must be established to be that of a responsible agent "eligible" for punishment. The second condition is related to the court's sentence: the punishment must find "its proportion" to the criminal act, establishing the right price to be paid to society. Given that both of these conditions were met by the Singleton case at the time of the sentence, what further purpose would be served by enforced treatment of this death-row inmate? Ultimately, as pointed out by Hart himself,[20] the general possession of the capacities of understanding is a condition of the efficacy of law. For it is only in their presence that the state can presume to communicate to its citizens the orders, commands, or other rules or principles upon which rests the existence of law. However, as we will see below, cases like that of Singleton should not affect the efficacy of law as the understanding of facts is fully available to the agent at the moment of sentence, him or her having been defined as fully aware of the legal consequences of their actions at the time of the crime. In this light, we could perhaps seek the answer in a more political dimension. Authorities want to ensure that no-one can escape the payment of their crimes to society, even if they develop mental illness after having been sentenced. But, as Mitchell Berman points out in her article "Punishment and Justification",[21] we should draw a distinction between punishment and suffering. Suffering can be acceptable in the given punishment if and only if the suffering is not seen as intrinsically good, but as something that the wrongdoer deserves. It would therefore not be morally justified to enforce medical treatment on a non-consenting prisoner on the basis that his mental state should be restored in order for him to suffer "competently" the result of his actions. So, is a minimal level of competence needed by society in order to ensure that the prisoner is granted all of his rights until the very last minute, or do we instead want to make sure that the inmate's suffering during his final moments is fully felt and perceived? Before answering these questions, the role of insanity in law will be considered in the next sub-section.

---

[19]Hart (1968), p. 160.

[20]Ibid., p. 229.

[21]Berman (2008), pp. 272–273.

3.5.2 The Evolution of the Role of Insanity in Law

In the development of the role of insanity in Anglo-American legal systems, the establishment of the full competence of the moral agent at the time of a criminal action has gradually become more relevant to establishing the level—or absence—of responsibility.

> To establish a defense on the ground of insanity, it must clearly be proved that, at the time of the committing of the act, the party accused was laboring under such a defect of reason, from disease of the mind, as not to know the nature and quality of the act he was doing; or if he did know it, that he did not know what he was doing was wrong.[22]

Even if the history of recognising mental conditions is much longer,[23] the so-called M'Naghten Rule of 1843 established the standard for the insanity defence. It did so by considering the mental capability relevant for the assessment of the *mens rea* in a criminal act. Central to its relevance was the idea that if a person is mentally ill and unable to distinguish between right and wrong, for example, then he or she cannot be held criminally culpable in our society. The principle upholds human dignity and ensures that those individuals acting against the law without malicious intent—such as people with severe delusions—will not be unfairly punished. This rule, however, was criticised for being too rigid, since it allows only severely mentally ill agents to be excused for their criminal conduct. In order to make it more flexible, in 1886 the decision in the Parsons v Alabama[24] case introduced some additional criteria for insanity defence. The court decided that a person could appeal by defence of insanity if she could prove through the application of what became known as the "Irresistible Impulse Test" that:

> by reason of duress of mental disease he had so far lost the power to choose between right and wrong, and to avoid doing the act in question, as that his free agency was at the time destroyed.[25]

The justificatory presumption would be that no matter the circumstances, for instance even in front of a police officer, the individual would not refrain from acting in the prohibited manner thus proving his or her lack of control. Subsequent cases further underlined control as an essential element of the *mens rea*[26] until 1970 when the American Law Institute introduced the Model Penal Code (MPC) with the intention of solving the increasing number of controversial cases related to the mental state of agents involved in criminal acts. The MPC denied responsibility of the agent involved in a criminal offence if:

---

[22]M'Naghten's case (1843b).

[23]Eigen (2004), pp. 395–412.

[24]*Parsons v Alabama*, 81 AL 577, So 854 1886 AL.

[25]Ibid.

[26]See, amongst others: *Sinclair v State of Mississippi,* 132 So. 581 1931 MS, *State v Strasburg*, 110 P. 1020 1910 WA, *Leland v Oregon*, 343 US 790 1952 OR, *Durham v United States*, 214 F.2d 862.

at the time of such conduct as a result of a mental disease or defect, he lacks substantial capacity either to appreciate the criminality of his conduct or to conform his conduct to the requirements of the law.[27]

Having considered the evolution in Anglo-American law of the role of the agent's mental sanity prior to the sentence, we can now move on to the next sub-section, in which we will combine all the elements considered thus far in a more exhaustive analysis of the controversial justification given in the Singleton case to enforce treatment on a person whose mental state deteriorated after the death sentence was passed.

3.5.3 A Retributivist Argument

In his article "Between Madness and Death: The Medicate-to-Execute Controversy",[28] Barry Latzer considers the Singleton case in detail, arguing that the decision of the Eighth Circuit Court of Appeals to forcibly medicate in order to carry out the execution procedure was a constitutional decision and, above all, a morally sound one. In other words, Latzer suggests with no hesitation that the court decision was in line with all the principles and directives highlighted in the course of this work. In this sub-section, I will explain his position in greater detail so as to then criticise its reading of the facts and propose an alternative solution in the subsequent parts of the chapter.

Latzer's reasons concern the state's need to legitimise its role as administrator of justice and to avoid exposing the system to exceptions to the retributivist principles at the very base of our current legal framework. Latzer proposes the following three policy options to deal with an inmate sentenced to death, suffering from mental illness, and potentially treatable such that competence to carry out the sentence could be achieved:

A *Medicate and Execute.*
The state carries out the standard procedure after having forcibly medicated the inmate and restored the minimal level of competence necessary.

B *Don't Medicate, Don't Execute.*
The execution of the death sentence has to be postponed indefinitely until the competence is restored either by unforeseeable factors (such as unusually positive developments of schizophrenia, dementia, etc.) or by autonomous decision by the prisoner to undergo treatment.

C *Medicate, Don't Execute.*
The state "bargains" for enforced treatment by downgrading the sentence to non-capital punishment.

---

[27]Model Penal Code (1985), p. 95.

[28]Latzer (2003), pp. 3–14.

On the one hand, as Latzer rightly points out, although option C might seem more humane at first glance, the use of such an approach would represent an injustice towards all those prisoners not developing any mental illness after having been sentenced to death. Due to a lack of space this problem will not be investigated further in this chapter, but surely the enormous discrepancy between sane and insane prisoners sentenced to death cannot allow us to consider option C as a morally justified and logically sound approach to future cases similar to that of Singleton. Concerning option B on the other hand, it might be argued that executing the prisoner without prior medication could be as cruel as rendering him competent at the time of execution, since he would thus live in an appalling mental state until the full capital punishment procedure was carried out. In some ways, Latzer supports this view by suggesting that option B is "unacceptably cruel".

In this light, we could say that not curing the prisoner and letting him live would be worse than curing and killing him. It would follow that if we could find a comparable punishment—for example, 10 years of imprisonment without medication prior to execution—to ensure respect for the retributivist principles needed by society, we should apply it without adding an extra punishment to those developing mental illness in prison. This aspect would surely represent an unfair addition to the suffering of already unfortunate individuals, making it difficult to defend as morally justifiable. However, a reliable way of assessing such a punishment is not currently available, and therefore the "readjustment approach" of option B is not possible, as the full avoidance of the capital punishment because of mental illness would also produce an unfair asymmetry between sane and insane inmates sentenced to death. Option A, Latzer affirms, is in truth the most convincing and consistent way of dealing with controversial cases like that of Singleton, as it ensures respect for retributivism—which is lacking in option C—as well as for the dignity of the individual—which is not guaranteed by option B. However, as Lawrence Gostin rightly points out:

> The Court holds that compulsory treatment must be medically 'appropriate,' but what if treatment will lead—directly or indirectly—to capital punishment? [...] The treatment would, at best, alleviate a patient's symptoms, but only in order to achieve a distinctly non-therapeutic end, namely, execution.[29]

Before proceeding in proposing a different option to those listed by Latzer in the final part of the chapter, I shall underline the importance of the link between the moral and legal justification, fundamental if we want to have a stable and consistent approach to what is morally sound. I will do so by taking into account cases that have functioned as cornerstones for the establishment of what is morally—and therefore legally- permissible when mental conditions are at stake. These cases were perhaps considered insufficiently by Latzer.

[29]Gostin (2003), p. 12.

3.6 Right to Treatment or Duty to Be Treated?

> In Anglo-American law, the criteria of criminal responsibility converge with the criteria of moral responsibility: where moral claims are warranted, so generally is legal sanction; and where there is moral excuse or justification, so too there is legal excuse or justification.[30]

However, it is important to underline that, as Christopher Kutz correctly emphasises, the key aspect is whether the forcible administration of drugs in the Singleton case had moral justification or not, for if it did not, its legal justification would be undermined too. In the decision process undertaken by the Eighth Circuit Court of Appeals various landmark cases concerning prisoners affected by mental illness were considered. The first of these was the 1986 Ford v Wainwright case, in which the Supreme Court ruled that the possibility of executing the insane was implicitly prohibited by the Eighth Amendment against cruel and unusual punishment. Judge Powell Jr. stated that: "the Eight Amendment forbids the execution only of those who are unaware of the punishment they are about to suffer and why they are to suffer it."[31]

This case became so relevant for the evaluation of the level of the mental capacity that it is now standard procedure to assess the eligibility for execution of prisoners on death row in relation to their "Ford competence". For the Singleton case, it is important to note that such competence refers to the prisoner's level of competence at the time of the execution rather than at the time of sentencing or of perpetration of the crime. This aspect refers to the issue of personal identity already considered, which requires further clarification at this stage. The CST to which we referred in Sect. 3.3 was certainly present, but a possible objection is raised by the fact that the diminishment of mental capacity during incarceration resulted in a situation in which the Competence to Be Sentenced (CBS) was lost. As with the Singleton case, in the Ford case no one disputed the full mental capacity of the defendant at the time of the crime, the trial or the sentencing. But Singleton lost his CBS *after* the sentence and despite having been assigned a penalty in line with the standards of the legal system, it is undeniable that incompetence at the time of the execution created a problem as his mind had by then deteriorated sufficiently to require a reconsideration of the original sentence. Clearly, the latter part of Judge Powell Jr.'s statement would conflict with the execution of a prisoner in a debilitated state of mind, as it would be difficult to consider him capable of understanding why he would be executed at the time of the actual execution. However, as seen above, the justification behind the continuation of the criminal punishment procedure would lie in the fact that, at the time of the initial sentence, the prisoner did understand the reasons behind his execution, thus making Singleton eligible for capital punishment.

As explained in earlier sections of the chapter, the a priori presumption is that there is continuity of the agent's personal identity over time. Under normal

---

[30] Kutz, C., Op.Cit, p. 571.

[31] *Ford v Wainwright* (1986).

circumstances, therefore, all the treatment intends to do is to re-establish that "normality" present at the time of the sentence and the crime. The precedent for enforcing treatment on a prisoner was thus the US Supreme Court decision in the Washington v Harper[32] case, which introduced the definition of "Harper involuntary medication". Some consideration of the differences between these related cases should be given, however: on the one hand, we should note that in the Ford case—unlike that of Singleton—the mental illness could not be eliminated entirely. In the Harper case, on the other hand, the main difference is that Harper was not on death row, and therefore curing him against his will would not have resulted in his death.

Also relevant is the 1989 case of Penry v Lynaugh,[33] in which the Supreme Court stated that "it is not cruel and unusual to give the death penalty to mentally retarded criminals". Juries, however, must be allowed to decide whether defendants should be given a prison sentence instead of the death penalty in light of their mental impairment. We should observe, in relation to the M'Naghten Rule considered above, that the judgement in the Penry case represents a significant change in the consideration of liability for individuals with mental impairment, deeming them as eligible for the death penalty as anybody else. Only the *actus reus*, and not the *mens rea*, is considered.

If that is the case, however, it would be fairer to assess the punishment in accordance with such a scheme. In other words, if the evaluation of the culpability of the agent does not take into account the *mens rea*, why should the resulting punishment be directed towards a restoration of a mental state that did not figure in the equation that led to the passing of the sentence in the first place? In the last section of this chapter, I will propose a solution that seems more coherent with the principles and cases involved in the judgements above.

3.7 A Further Option

In this final part of the chapter, I will briefly recapitulate the key points highlighted above, as this will provide the theoretical basis by which to understand the claim that this chapter aims to make.

At the beginning of this chapter, I listed three conditions that would guarantee the "fairness" of forcing medication on a prisoner, and I concluded that at least two of them are not applicable in the Singleton case. Through the analysis of landmark cases, I then considered possible alternative justifications for the treatment that would still be consistent with the standard approach to both punishment and mental illness. The final part of Sect. 3.4 considered Latzer's synbook, wherein the legal and moral dilemmas of the Singleton case can only be resolved by medication then

[32] *Washington v Harper*, 494 US 210.

[33] *Penry v Lynaugh*, 492 US 302.

execution, as this is the only way of respecting both retributivist principles and the dignity of the individual.

In Sect. 3.5 however, I have questioned this position by upholding that, even if it may be constitutionally sound, enforced medical treatment is less morally justified than Latzer affirms. Evidently, my critique is based on a very different interpretation from Latzer's of how satisfactorily the precedential judgements underpinning the Singleton case were utilised.

Hence, if the key aspect of restoring competence before execution is to be consistent with retributivist principles, I could accept—as Latzer does—his option A as morally sound if and only if the authorities involved—the state as well as the more specific court issuing the sentence—were ready to clearly affirm that the motivation for re-establishing competence is simply to ensure the well-being of society and the continuity of its rules and legislative system. Currently, however, as shown in the Singleton case, the restoration of competence is presented as something in the best interest of the patient, and not of society. This inconsistency between the need for strictness and at the same time an unwillingness to publicly admit the real values at stake in such a decision makes option A unconvincing.

In order to overcome this deadlock, I will propose one additional option on top of those provided by Latzer:

D *Don't Medicate, Execute.*
The death penalty procedure should be carried out without taking into account the lack of competence the patient might be temporarily or more permanently suffering from.

This alternative might appear inhumane at first, but of the four options available within the current Anglo-American legal system it is possibly fairer for both the inmate and the state. Differently from the other options, my formulation would for two reasons apply more consistently the various principles put forward in the analysis of the Singleton case. Firstly, it would not distort the interpretation of such principles in one sense or the other according to convenience. Secondly, it would also make a more convincing use of the cases used in the deliberation of the Singleton judgement.

Regarding the first point, I would claim that despite accepting the need for the retributivist principles to be respected, we have acknowledged in the course of this chapter that the restoration of competence in an individual involves many contested issues still far from being definitively resolved. After all, there is not yet even agreement on whether to consider the agent in question fully retraceable to the perpetrator of the original crime! If we want to guarantee justice and fairness in such a nebulous context, we need to be sure that we are not provoking unnecessary harm. Given that enforcing medical treatment could well be seen as an avoidable harm, we might be better off avoiding its implementation for the time being. By allowing for the execution to take place, however, this option would ensure that the retributivist principles prevalent in society would still be served.

In relation to the second point, in the previous section key legal cases were introduced in order to re-examine the more empirical frame of reference used by the judges. We underlined that both the Ford case and the Harper Case played an important role in the passing of the sentence. Their relevance to the Singleton case, however, is not unquestionable, as differences from Singleton's situation existed in each of them. The application of option D would avoid inconsistent interpretation of the Ford and Harper verdicts when applied to relatively similar cases. Without taking away the inherent value of achieving a more stable usage of those cases, yet more important is the fact that my proposed alternative would deal with the distinction between the *actus reus* and the *mens rea* in a more clear-cut manner. A coherent application of the Penry case would be guaranteed throughout the whole process, including the capital punishment procedure itself. If the *actus reus* is the only factor that counts in the equation when assessing the punishment, so it should be accepted that there would be no more—moral or legal—justification to restore the mental conditions presumed by *mens rea* not previously taken into account in establishing guilt. By executing without forcibly curing the prisoner this very controversial aspect of the Singleton case could be avoided in similar future situations.

3.8 Conclusion

In conclusion, I have tried to show that within the current Anglo-American legal system the justification for forcible medical treatment of death-row inmates is difficult to defend on either moral or legal grounds. On the one hand, the impossibility of changing the prisoner's sentence makes it problematic to claim that involuntary drugging would represent a better option for him. On the other hand, the will to re-establish competence in the patient in order to ensure that his suffering is fully proportionate to the crime committed is difficult to accept.

To broaden the perspective on how we should deal with future cases similar to that of Singleton, a historical analysis of the idea of punishment in the Western tradition was taken into consideration, with a special emphasis on the evolution of the role of insanity in law. Subsequently, to highlight the diverse interpretations that insanity has had in different contexts, various landmark cases were analysed, giving rise to important questions about the consistent application of the principles and justifications underpinning their final judgements. In order to critique the current acceptance of the position evident in the Singleton case, I employed Barry Latzer's influential work, ultimately going beyond it in proposing an additional way out of the Singleton quandary.

In considering the three options suggested by Latzer, I agreed that option C would certainly be the most tempting from a "humanitarian" point of view as this option would have the law allow a possible readjustment of the death sentence after the enforced medical treatment (perhaps to lifelong imprisonment). In this way, the claim that enforced treatment serves the best interest of the patient could be justified, but this possibility naturally flags up the related issue of the inequality of treatment between prisoners who develop mental illness while on death row and those who do not. This would leave competent death-row inmates paradoxically hoping to develop some kind of mental illness in order to avoid capital punishment. As a result, I concluded that all three options, including the one (option A) used by Latzer to legitimise the legal and moral acceptability of the decision taken in the Singleton case, are indeed unsatisfactory on both moral and legal grounds.

Synbooking the analysis carried out in this chapter, I conclude that the only reason for such a treatment would be based on its political value, and the need to re-establish competence is only related to the desire of the relevant authorities not to allow a "soft message" to filter out from this case. I do not aim to question the acceptability of such a justification here, but it should become apparent at this stage that the enforcement of treatment has been based on fictional principles such as consideration of the prisoner's best interest. This distinction between the hidden message of the sentence and its "politically correct" version is what, in my opinion, makes its moral foundations inevitably unstable. As a result, to support this kind of approach in future cases similar to that of Singleton seems unjustifiable.

Rather, to avoid the continuation of such injustices in future, I suggested a new approach to cases resembling that of Singleton. I argued that it would be more coherent to hold that, once the agent is established to have been indisputably competent at the time of the death sentence, the authorities should continue with the capital punishment procedure without any further hesitation related to the mental condition of the prisoner.

Alternatively—should this option be regarded as inhumane—we would have to find a new and more consistent way of dealing with cases involving autonomy and competence. Before doing that, however, in the next part of the book a contrasting use of enforced treatment will be considered: namely, its use to keep a person alive even when they are considered to be competent.

References

Berman, M.N. 2008. Punishment and justification. *Ethics* 118: 258–290.

Douglas, T. 2008. Moral enhancement. *Journal of Applied Philosophy* 25(3): 228–245.

Duff, R.A. 1998. Responsibility. In *Routledge encyclopedia of philosophy*, ed. Craig, E.J. London: Routledge.

Eigen, J.P. 2004. Delusion's odyssey: Charting the course of Victorian forensic psychiatry. *International Journal of Law and Psychiatry*, 27: 395–412.

Ford v Wainwright. 1986. 477 US 422 (Justice Powell, concurring).

Gerbasi, J.B., and C.L. Scott. 2004. Sell v. U.S.: Involuntary medication to restore trial Competency-A workable standard? *The Journal of the American Academy of Psychiatry and the Law* 32(1): 83–90.

Gostin, L. 2003. Compulsory medical treatment: The limits of bodily integrity. *The Hastings Center Report*, Sep–Oct 2003, p. 12.

Gross, M.L. 2005. Treating competent patients by force: The limits and lessons of Israel's Patient's Rights Act. *Journal of Medical Ethics* 31: 29–34.

Hart, H.L.A. 1968. *Punishment and responsibility.* Oxford: Clarendon Press.
Hope, T. 1994. Personal identity and psychiatric illness. In *Philosophy, Psychology and Psychiatry*, ed. Griffiths, P.A. Cambridge: Cambridge University Press.
Kutz, C. 2004. Responsibility. In *Jurisprudence and philosophy of law*, eds. Coleman, J. and Shapiro, S. Oxford: Oxford university Press.
Ladds, B. 1995. Involuntary electro-convulsive therapy to restore competency to stand trial: A five year study in New York State. *Journal of Forensic Sciences* 40(2): 183–187.
Latzer, B. 2003. Between madness and death: The medicate-to-execute controversy. *Criminal Justice Ethics* 22(2): 3–14.
Model Penal Code. 1985. American Law Institute, Philadelphia, 2007. In *Crime and madness*, ed. Maeder, T. New York: Harper and Row Publishers.
M'Naghten's case. 1843b. UKHL J16 (19 June 1843). http://www.bailii.org/uk/cases/UKHL/1843/J16.html. Accessed 4 Jan 2015.
Peloso, P.F., and T. Bandini. 2007. Follia e Reato nella Storia della Psichiatria. Osservazioni Storiche sul rapporto tra Assistenza Psichiatrica e carcere. Rassegna Italiana di Criminologia, Anno 1, 2, Pensa Multimedia Editore.
Parfit, D. 1984. *Reasons and Persons.* Oxford: Clarendon Press.
Vincent, N.A. 2009. What do you mean i should take responsibility for my own ill health? *Journal of Applied Ethics and Philosophy* 1: 39–51.
Wolf, S. 2003. Sanity and the metaphysics of responsibility. In *Free will*, ed. Watson, G. Oxford: Oxford university Press.

Chapter 4
Hunger Strikes and Other Controversial Cases

4.1 Introduction

In order to substantiate the claim made in Chap. 3, the attention of this book will now shift towards a further two controversial cases relating to the [mis]use of the notion of autonomy. The first case relates to the forced treatment of a burns victim desirous of death, and despite dating back nearly three decades, it remains very topical, raising important questions pertinent to the current study. Indeed, the relevance of this case is such that it is amongst the most frequently examined in bioethics courses at US institutions.[1]

The second case is rather more recent and focuses instead on the absence of forced treatment of a hunger striker in Italy. Given that this project aims to provide an accurate perspective of autonomy in Western contexts beyond strictly Anglo-American boundaries, the geopolitical element of this case constitutes an additional reason for including it in the work. Compared to other issues within the European Union (EU), Hunger Strikes have been of relatively minor importance. However, a recent case occurring in Italy has focused attention on the issue, underlining a general uncertainty within the EU with regards to the topic and suggesting that a more firm and consistent standpoint is required.

Amongst the member-countries of the EU there is still little clarity over the approach that the law should take towards respect for patient autonomy. There are a number of reasons for this: first of all, approaches to the notion of autonomy can differ substantially if tackled against a more secular or more religious backdrop. These differences are noticeable in many contexts, and they surely represent an interesting theme worthy of investigation. In this chapter, however, the focus will be directed instead towards a specific representative of the [more religious] southern member-states of the EU, namely Italy, and the application of the principle of autonomy within that context.

[1]Given the extent of literature produced on this case, this book can only pay attention to some specific aspects of the numerous controversial issues raised by it.

M.D. Garasic, *Guantanamo and Other Cases of Enforced Medical Treatment*, SpringerBriefs in Ethics, DOI 10.1007/978-3-319-22653-8_4

The reason why these two cases figure within the same chapter is simple: differently from the two situations considered previously, in both of the cases foregrounded in this chapter the presence of a mental illness is far from given. Nonetheless, strong claims are made as to the *temporary* competence of the people involved in them. What is particularly interesting when comparing these relatively similar circumstances is, once again, the absence of uniformity in affirming when and how a person is autonomous -or, to use a term more strictly related to the psychiatric dimension: competent. The current analysis aims to raise additional doubts as to the appropriateness of disparities of treatment justified in the name of the same notion of autonomy.

4.2 The Dax Case

On the 23 July 1973, Donald "Dax" Cowart's life changed beyond all recognition. Due to a gas leak and a series of unfortunate events, he and his father remained trapped in an inferno caused by a propane explosion. They were both brought to a local casualty unit, but his father died on the way to the hospital. Donald Cowart's life was saved due to the extensive and painful treatment that he received explicitly against his will. He consistently expressed his desire not to continue his life as he was aware that what was awaiting him was going to be unbearably different from the life he was accustomed to.

Before the accident, Donald was a young man full of energy, with a great sense of independence and with good prospects on both the professional and the sentimental fronts. He was weighing up his career options; whether to continue training to be a pilot or to finish law school and join his father's business. He had also recently started a relationship with a young woman. Following the accident, Donald knew that he was bound to be dependent on other people from that moment on and did not want to continue living.

However, despite his insistence on being allowed to die, the team of doctors in charge -morally pushed to continue by Donald's mother and legally uncertain about their potential liability-[2] decided to override the express wishes of this competent patient and continue medical treatment. The justificatory principle used at the time was that of *soft paternalism*: once Donald had reached the other side of this painful journey and come to appreciate -even if in a more limited and certainly different way- life again, he would retrospectively agree that it had been the right decision to take.[3]

[2]For the sake of intellectual honesty it must be underlined that at that time the Texan approach to allowing competent patients to refuse treatment had just changed in favour of a less permissive attitude towards the patient's will. Donald Cowart was particularly unfortunate in this respect also.

[3]In relation to this point see the three conditions provided by Michael Gross and listed in Chap. 3.

Despite losing his sight and the use of both hands, he eventually left hospital, managing to get a law degree and become an attorney. He changed his name to Dax[4] and married (although he is now divorced). He also successfully sued the company liable for the accident that took his father's life and part of his own.

The fact that Dax Cowart managed to finish law school, get married and run a relatively successful company might lead us to think that forcing medical treatment upon him against his will was indeed the right choice to be made. This initial impression, however, should soon be challenged alongside the moral justification underlying this undeniably paternalistic approach to the Dax case.

The first point that should be taken into account is that, in order to justify treatment, his mental state, his competence to make a decision over his own body and life were called into question, infringing in the most direct way a Millian standard of non-interference with the liberty of another individual. Even if an argument for a *temporary* lack of autonomy could be made,[5] it would be indefensible to claim that we could allow this exception to cover the entire period during which Dax had to suffer the treatments and operations as this lasted nearly 10 years.

The reason for needing to prove the absence of competence at the moment of the decision to refuse treatment was based on the standard procedure used in the US. As correctly summarised by Tristam Engelhardt in his commentary on the Dax case:

> When the patient who is able to give free consent does not, the moral issue is over. [...] In short, one must be willing, as a price for recognizing the freedom of others, to live with the consequences of that freedom: some persons will make choices that they would regret were they to live longer. But humans are not only free beings, but temporal beings, and the freedom that is actual is that of the present. Competent adults should be allowed to make tragic decisions, if nowhere else, at least concerning what quality of life justifies the pain and suffering of continued living. It is not medicine's responsibility to prevent tragedies by denying freedom, for that would be the greater tragedy.[6]

In his analysis of the Dax case, Engelhardt suggests -on top of the vital importance of respecting a competent patient's decision- an additional aspect that in his opinion deserves attention: time. We are after all temporal beings and what makes the difference in the way we live life are the choices that we make in the present. However, as already mentioned in Chap. 3, for reasons of space this work cannot enter into the debate over the continuity of the agent in time.

Nonetheless -and this is the second point- it might be interesting to consider that in Dax's particular case, even the future Dax was going to be against the treatment. Contrary to standard expectations, in fact, Dax has consistently claimed that he should have been allowed to die, even when his quality of life returned to a

---

[4]There are different interpretations of the reason for this choice, but what is common to all of them is that the name change was directly related to the accident.

[5]When describing the Dax case (referred as Mr. G), Robert Burt writes: *"It may seem that some new label should be devised to categorize Mr. G's confusion -perhaps 'temporary incapacity'— that would not tar him with the mental illness brush."*(Burt 1979, p. 13).

[6]Engelhardt (1989 p. 96).

tolerable level. That is, after 10 years of forced treatment.[7] It follows that the condition of the principle of proportionality that allowed forced treatment would fail in this case. Regarding the definition of such a principle in this context, the description provided by Albert Jonsen can be useful:

> No form of treatment, such as nutrition and hydration or resuscitation or antibiotics, can be considered universally warranted or obligatory. This conclusion is sometimes described as the principle of proportionality, in which an assessment of the proportion of benefits to burdens, as evaluated by patients, physicians, and families, dictates the ethical conclusion.[8,9]

It seems clear that in the deliberation process for Dax, the first of these elements - namely patient Dax- was not considered at all.

What is most striking in this case, from Dax's point of view, is that despite having been found to be competent enough to make decisions over his own life -a substantial difference from the situations highlighted in Chaps. 2 and 3, the enforcers of his treatment decided that his choice not to continue with medical procedures was to be ignored. In other words, Dax was not allowed to freely shape his destiny, nor was he allowed to die, even though this is what would have happened if treatment had not been provided from the beginning of this dramatic story. This was because the authorities in question -doctors and family- chose another destiny for him: the doctors involved were clearly also worried about the legal consequences of their actions, while the family members -especially his mother- were moved by good intentions, but probably also by a form of selfishness in not wanting to let him go. Ultimately, everyone was satisfied with the decision except the one suffering its consequences.

Dax Cowart explicitly decided to put an end to his life from the very beginning of -what would have then become- his second life. He asked the first person coming to his aid following the explosion to give him a gun with which to shoot himself; then he told ambulance staff that he did not want to be kept alive; and finally, once at hospital he clearly stated that he did not want to undergo treatment, a position he sustained well into the advanced stages of his incredibly painful and challenging recovery process.

---

[7]Concern for Dying: Dax's Case, videotape, 1985, Unicorn Media, New York.

[8]The words "proportionate" and "disproportionate" are not in the text of the report, but they do appear in a footnote on the same page as the abovementioned conclusion, in a citation from the Vatican's Declaration on Euthanasia (June 26, 1980) which is reprinted in an appendix to the report. While there are subtle differences between the Vatican's concept of proportionate care and the Commission's conclusion, there is a common theme: a treatment is not morally obligatory when, in the patient's view, it produces greater burdens than benefits. See President's Commission on Ethical Problems in Medicine and Biomedical and Behavioral Research, Defining Death: A Report on the Medical, Legal, and Ethical Issues in the Definition of Death (Washington, D.C.: U.S. Government Printing Office, 1981), p. 89 as quoted in Jonsen (2003, p. 260).

[9]Jonsen (2003, p. 260).

It is important to notice that during an interview for his film "Please Let Me Die",[10] Dax shows an incredible sensitivity towards the positions in which all those involved found themselves, providing for a profoundly compassionate reading of the motivations of the individuals concerned. Nonetheless, while there can be no debate over the moral propriety of the first man on the scene or of the paramedics (neither actor being allowed to evaluate the patient's competence for the practical reason of needing to focus on saving her life), the same cannot be said of the doctors -and their power- involved in the case.

In their case, the choice not to heed Dax's will was perpetuated over time, so the only way to justify this long-term treatment was to jeopardise not only Dax's competence -a point stretched to the maximum but bound to fail- but also his very values. This pre-selection of what constitutes an "acceptable value" seems to be strongly illiberal and conflicts with the very idea of respecting one's autonomy as long as one remains competent.

Once again this inconsistent way of dealing with difficult cases was made possible by yet another redefinition of how exactly we are to respect a person's autonomy. In the Dax case it appears that the preferred way of understanding autonomy was to see it -very conveniently- as freedom of action.[11] This is the third point to consider: what was called into question, in fact, were Dax's values. As rightly pointed out by Richard Zaner:

> How could it possibly happen, Dax constantly implored, that in a society such as ours, whose moral focus is so firmly set in the right of the individual to determine his or her own course in life, precisely that right could *at the same time* be denied? An adult who was declared clearly competent and thus a person who just as clearly ought not to be denied the right of self-determination, yet *just this was in fact denied*, and Dax was forced in the most literal way to undergo extraordinary and agonizing treatments against his own specific and declared wishes. Massively compromised in bodily abilities, profoundly and permanently disfigured, he was made to face a future devoid of everything *he* valued.[12]

To better understand the meaning of such a statement, it is helpful to consider Robert White's analysis of the Dax case: "I think we shall never know whether Dax wholeheartedly wanted to die. But he demanded to die, and that was the issue that had to be dealt with at the time."[13]

The point highlighted by this quote is crucial: we can never be absolutely certain about the "true" intentions of a person when she makes a decision, not even if we are that person. The "wholeheartedness" itself hints at the inherent vagueness of such assessments, taking as it does the heart -a universal symbol of irrationality and unpredictability- as its point of reference.

Can we assert with any certainty that Mr X was fully competent and sufficiently informed to give his (unwavering) consent via his signing of a new mobile phone

---

[10]"Please Let Me Die", transcript of videotape available in Areen et al. (1984).

[11]See footnote 17 below for a more specific reference.

[12]Zaner (1989, pp. 43–44). My emphasis.

[13]White (1989, p. 18).

contract? Would we stop the procedure until that certainty could be reached? No, we would not, and this attitude would be based on the accepted practice and presumption that there is no way to achieve such a level of certainty. So, in order to prevent the entire market from grinding to a halt, we allow for the fact that this degree of inherent uncertainty is embedded in the very nature of human beings and continues to exist in the negotiation of a new mobile phone contract.

Here we can clearly see that the problem is not the level of certainty that we can have over genuinely autonomous individual choices but, rather, the question of what constitutes an acceptable choice. We do, in fact, have a tendency to consider this behaviour in line with one's predictable way of interacting with society. This is because the way of affirming one's autonomy (understood here as authenticity)[14] converges with the values that our society expects to be accepted. There are no grounds to affirm or suppose that Dax was not acting in line with the "linear autonomy" of his own values. What did not work in favour of his decision was the fact that his values were considered unworthy of respect, an approach that is the epitome of paternalism and has nothing to do with the defence of autonomy.

In relation to this point, Robert Burt's interesting analysis points out that refusing treatment could be a way of paradoxically reaffirming that same version of individual autonomy so strongly defended by modern Western society. Regarding the clash between the individual and societal values, he instead writes:

> Dr. White was initially brought to interview Mr. G by physicians who asked whether Mr. G. might be diagnosed as mentally ill so that the state civil commitment laws could be invoked to force treatment on him regardless of his consent.[15]

To conclude the analysis of the Dax case in relation to the present investigation two interesting parallels with the cases presented in Chaps. 2 and 3 can be drawn. These underline a disturbingly systematised incoherence in the application of the concept of autonomy.

The first point relates to Anorexia Nervosa and underlines the particular difficulties that we have in accepting the importance of the role that a patient's life-expectancy and life-quality play in our judgements relating to refusal of treatment. As James Childress and Courtney Campbell wrote:

> Please Let Me Die [...] effectively challenges viewers to consider how they would balance the principles of respect for persons and patient benefit when a patient refuses life-prolonging treatment even though he or she is not terminally ill, i.e. is not irreversibly and imminently dying, and life could be prolonged indefinitely with reasonably good quality.[16]

The second point that deserves attention, again from Childress and Campbell, is that amongst the versions of autonomy that could have been applied to the Dax case, he: "was not autonomous in the sense of free action, in contrast to effective deliberation,

[14]Here I refer to the definition of autonomy given by Ronald Dworkin and highlighted in Chap. 1.

[15]Burt Op. Cit., p. 3.

[16]Childress and Campbell (1989, p. 23).

it might be argued that continuation of treatment was essential to restore his autonomy as free action."[17]

This consideration has challenging inputs in both of the previously analysed cases. In relation to Anorexia Nervosa, this point is particularly interesting if evaluated in light of the percentage of suicides recorded after enforced treatment takes place. In the Singleton case, meanwhile, this freedom of action appears to be more related to his executioners rather than to him, as he would stay in prison anyway, with a very limited form of freedom of action.

After having considered some of controversial points raised by the Dax case, in the remaining sections of the chapter, I will move the attention to a more recent case of hunger strike where the level of competence of the patients in question was considered sufficient not to enforce treatment. In analysing this divergent way of respecting autonomy, I will try to establish the extent to which this is due to a "cultural evolution" in the application of the law, or rather down to pure convenience for the authorities involved.

4.3 Issues Related to Keeping One's Alive Against His Will

As noted in previous chapters, the supremacy of autonomy among other principles in the field of biomedical ethics has in recent years come under increasing challenge from many quarters[18] as an unsatisfactory base from which the law hopes to affirm its legitimate *super partes* role. In focusing on a recent case in which respect for autonomy was used as the central justification for the approach used by authorities in dealing with a controversial situation, the intention of this part of the chapter is to provide a linear comparison with the Dax case described above, sensitising the audience to the inconsistency of the application of the principle of autonomy and proposing that an alternative is required in order to guarantee fairer treatment in future.

To highlight this inconsistency, I will now put forward a critical -and more focused- analysis of the biopolitical application of the law. The utilisation of "biopolitical" terminology implies a reference to Foucault, the intention being to underline the lack of fairness in the application of the law in bioethical cases. The application of principles that should be coherently employed in a particular case should not be subject to the contextual weight of their decision. This is what occurred for Dax, and this is indeed the case in most situations where authorities have the ultimate decisional power over one's life.

---

[17] Ibid., p. 32.

[18] See, amongst others: Foster (2009), Gillon (2003, pp. 307–312), O'Neill (2002) and Harris (1994).

Through an analysis based on relatively similar cases worldwide, a reading of the aforementioned Italian example will be offered, putting forward three biopolitical reasons for the non-interventionist attitude evident in the cases described below.

4.4 Allowing to Die: The Case of Sami Mbarka Ben Garci[19]

In September 2009 a very controversial interpretation of the principle of autonomy developed in a prison in Pavia, Italy. Sami Mbarka Ben Garci, a Tunisian prisoner charged with rape died in his cell as a result of self-inflicted starvation. The reason for his hunger strike was related to the criminal charge itself: he denied being guilty of rape and, not being able to continue his life with such shame on his honour, he decided to slowly terminate his life as a form of protest against what he perceived to be an unfounded charge.

It would be impossible to establish convincingly Ben Garci's guilt or innocence, and out of respect towards both the deceased Ben Garci and the rape victim, this work will not question the court's verdict regarding the culpability of Ben Garci. Rather, my focus will be on attempting to establish the extent to which the social preconditions of an individual prior to a certain [criminal] action can change the value assigned to the life of a given human being in a particular (Italian) society. It will be argued that in Ben Garci's case this was indeed what happened, leading to a fierce defence of the principle of autonomy completely unprecedented in the Italian context. This lack of precedent in itself provokes many questions as to why tacit consent was given to Ben Garci in this particular case of hunger strike.

In fact, his autonomy was so deeply respected that no one decided to intervene forcibly even when his health was clearly deteriorating to a dangerous extent. After Ben Garci's death, the director of the prison claimed that: "to deprive someone already in prison of their self-determination power is cruel."[20] It seems obvious that such a "respectful" view of prisoners' autonomy is -fortunately- not the most common approach that prisons have towards their inmates, otherwise we might feel

---

[19]This part of the chapter draws extensively from a previous co-authored work of mine. I am grateful to both Charles Foster (the co-author of the article) and the editors of the journal for allowing the reproduction of the material. Garasic and Foster (2012, pp. 589–598).

[20]See, amongst others: La Repubblica, 9 September 2009, available at: http://ricerca.repubblica.it/repubblica/archivio/repubblica/2009/09/09/sciopero-della-fame-detenuto-muore-pavia.html [accessed on 18 January 2010], Il Corriere della Sera, 9 September 2009, available at: http://archiviostorico.corriere.it/2009/settembre/09/Pavia_morto_detenuto_tunisino_che_co_9_090909040.shtml [accessed on 4 January 2015], http://www.adnkronos.com/AKI/English/Security/?id=3.0.3746339292 [accessed on 4 January 2015], http://news.bbc.co.uk/2/hi/europe/8335092.stm [accessed on 4 January 2015].

ethically entitled not to intervene if we see a prisoner hanging himself without this producing any sense of guilt in us.

Ben Garci was admitted to hospital on 3 September. The next day his conditions, rather than improving, get worse. On 5 September, at 3:45 a.m., Sami Mbarka Ben Garci dies, leaving many questions over the culpability of a system not seeming to have convincingly tried to save his life. A few facts available exhibit a lack of clarity meriting our attention.

At the end of August 2009, prison doctor Pasquale Alecci decided to contact the relevant authorities about the deteriorating state of Ben Garci's health. The prisoner had given up eating solid food more than 40 days prior, and was at that point drinking only water with sugar. He lost 21 kilos and could hardly stand. He was, however, conscious and convinced of his choice, as it was his intention to actively protest against a penalty that he considered unjust.

With his condition deteriorating to a critical level, first the doctor, then the *Magistrato di Sorveglianza* (Surveillance Judge), asked the Ministry of Justice[21] to intervene by moving the prisoner to an institution capable of guaranteeing the appropriate care to the prisoner/patient. More specifically, they requested placement in an adequate *centro diagnostico terapeutico* (therapeutic diagnostic centre) with all the facilities necessary for the recovery of the inmates.

While waiting for clarification regarding the possibility of relocation, Ben Garci was moved on 1 September to an emergency hospital, because the quality of healthcare provision at the Torre del Gallo prison had declined so badly that neither a cardiologist nor a psychiatrist was available at the prison. When Ben Garci arrived at the hospital, he -in line with his longstanding position- refused any treatment. He was then visited by the psychiatrist of the hospital who found him fully competent and, therefore, ineligible for *trattamento forzato obbligatorio* (obligatory enforced treatment, TFO). Ben Garci was sent back to prison.

On 2 September the Ministry responded negatively to the Magistrato di Sorveglianza's request. The lack of justification for transferring the prisoner was based on the claim that Italy could not count on *centri clinici penitenziari* (penitential clinical centres) able to deal appropriately with Hunger Strikes. It was suggested, however, that close watch be kept over the prisoner with possible intervention via TFO. The same day, the mayor of Pavia signed documents permitting such treatment.

Also on 2 September, the magistrate involved in the case decided to override the Ministry's decision and arranged for the transfer of Ben Garci to a non-penitentiary institution (Policlinico San Matteo). He justified his decision by claiming to find himself in disagreement with the principles underpinning the Ministry's response: the objective of hospitalising the prisoner was not to cure him, but only to ensure that his medical condition did not become life-threatening.

[21]Available at: http://www.senato.it/japp/bgt/showdoc/frame.jsp?tipodoc=Resaula≤g=16&id=004312360=doc_dc-allegatob_ab-sezionetit_icrdrs&parse=no [accessed on 4 January 2015].

This is the key point of the whole issue: are we allowed to intervene in order to preserve the sanctity of life[22] beyond any other principle, namely that of autonomy? In other words, should the authorities have put aside an evaluation of competence in the prisoner in favour of a focus only on the gravity of his medical condition?

Given a positive responsive to these questions, the priority of saving life over any other principle would have been affirmed, but clearly -while this was true for Dax- this was not the case for Ben Garci. The next question that we must pose, then, is: was this due to the fact that Ben Garci was a foreign prisoner, an element that society does not care so much about? In the following sections we will attempt to provide an answer to this important question.

4.5 Can We Consider Reliable the Competence of a Hunger Striker?

Hunger Strikes produce a number of medico-ethical questions. We can schematise them as follows:

Step 1 Do we respect the person's will in the extreme or do we have to intervene when life is at stake?

Step 2 If we decide to intervene, does the preservation of life have such primacy as to allow force-feeding to take place no matter how invasive this procedure might be, or can we accept that there are cases where keeping a person alive will result in more suffering than letting him die?

Step 3 If we allow for the possibility that some forms of force-feeding entail excessive suffering, we will have to reconsider our whole approach to the sanctity of life as an absolute principle; the recognition of this would take us back to Step 1.

We shall explore Steps 2 and 3 later on in the work when considering additional hunger strike cases, but, in relation to Step 1, we must bear in mind the current definition of Hunger Strikes in law, as well as its dependence on the notions of autonomy and, more specifically, competence. A hunger striker has recently been characterised by the World Medical Association (WMA) as a mentally competent

[22]For the sake of the current discussion, the reader should consider a definition of the sanctity of life very much in line with that described by Ronald Dworkin (1993). However, when analysing the specific Italian context, it is appropriate to take into account the deep influence that Roman Catholicism still plays in bioethical controversies, with a resulting attitude that tends to see life as something to be preserved under any circumstances, no matter how extreme. However, we shall later discover that -due to their biopolitical value- some exceptions have been made to this normally intransigent way of portraying sanctity of life.

person "who has indicated that he has decided to refuse to take food and/or fluids for a significant interval."[23]

However, this loose definition should not lead us to believe that there is no connection between Hunger Strikes -being rooted in political or personal motives- and mental illness. On the contrary, there is substantial evidence that supports the idea that, even where initially competent, prisoners partaking in Hunger Strikes suffer from a multiplicity of mental disturbances related to their imprisonment.

As rightly pointed out by Mary Kenny, Derrick Silove and Zachary Steel[24] in a study on US-based asylum seekers' Hunger Strikes,[25] despite being fully competent and willing to commit to their cause (obtaining asylum) at the beginning of their protest, towards the end of their detention -especially in those cases where their asylum applications had already come back negative- the continuation of the Hunger Strikes was motivated not by a desire to protest but rather by a will to die in order to avoid the suffering of the extremely hostile situation they would face once repatriated. This awareness leads us to question the advisability of affirming the presence of competence in these situations.

Similarly, we should also note that some studies conducted on political hunger strikers in South Africa during the Apartheid regime found that levels of clinical depression amongst the prisoners involved were as high as 77 %.[26]

In relation to this issue, it is also interesting to consider the analysis carried out by Gürkan Koçan and Ahmet Öncü in relation to Hunger Strikes in Turkey.[27] The main focus of their research was an exploration of the political value of Hunger Strikes in the Turkish context, concluding that Hunger Strikes undertaken in Turkey have to be recognised not only as political battles that deserve respect for their meaning but also as competent choices made by citizens with regard to their own lives.

In considering the moral justifications underlying possible intervention by authorities -as has occurred in Turkey-[28] the authors adopt a Kantian standpoint on the issue, affirming that intervention denies the respect of individual autonomy and, as a result, dignity. Given the Dax case, this affirmation seems peculiar to say the least. It should be noted, however, that their way of interpreting Kant is open to counter critiques. For example -still using a Kantian approach- one could provide at least two arguments that would undermine such a justification of Hunger Strikes: first of all, even if dying is not the ultimate intention of the prisoner, death is foreseeable and therefore should be avoided, Kant coming out explicitly against

[23]World Medical Association. Declaration of Malta on Hunger Strikers (adopted by the 43rd World Medical Assembly in Malta, November 1991 and revised at the 44th World Medical Assembly in Marbella, Spain, November 1992). Geneva: WMA, 1992. The document has been revised again in the 57th General Assembly in Pilanesberg, South Africa, October 2006.

[24]Kenny et al. (2004, pp. 237–240).

[25]Brockman (1999, pp. 451–456).

[26]Kalk et al. (1993, pp. 391–394).

[27]Koçan and Öncü (2006, pp. 349–372).

[28]See, for example: http://news.bbc.co.uk/2/hi/europe/1722075.stm, and footnote 29 below.

suicide irrespective of the motive principles behind the act. Second, the utilisation of any human being as a tool or -to use more Kantian terminology- a means to an end, is not permissible. This principle should be applied even to one's own body, resulting in a condemnation of this self-inflicted death in service of a political message.

Remaining in a Turkish context, and still focusing on the questions raised above, it is interesting to note Murat Sevinç's point in his "Hunger Strikes in Turkey"[29] that we should acknowledge that Hunger Strikes have frequently been redefined as Death Fasts so as to stress the striker's awareness of the risk that his fast represents. The conscious choice not to avoid death for a higher cause -and the reaction of authorities to such a decision- will be considered more in depth in Sect. 4.8, but for the time being, suffice it to say that this form of protest often produces very harsh reactions from institutions of power, as described above.[30]

This consideration emphasises another important dimension of Hunger Strikes and Death Fasts. These phenomena represent in many contexts the only available "tool" with which to attempt to send out a political message. Not allowing for their continuation in the name of the respect for strikers' lives can thus be seen as an efficient way for the authorities to mitigate the political effects of the strike. This certainly is a limiting factor for an individual's right to choose competently what is best for him.

Given the difficulty of reaching a consensual agreement on the definition of autonomy to be applied in cases of hunger strike, the next section will examine the more medico-legal definition of autonomy and rationality -namely competence- as well as the role it plays in decisional processes concerning Hunger Strikes.

As a result of the considerations outlined above, could we convincingly affirm that the choice of the prisoner is not affected by external events? The position of this work would suggest that we should not. After all, we should not forget Anita Ho's point that: "discussions of patient autonomy in the bioethics literature, which focus on individual patients making particular decisions, neglect the social structure within which health-care decisions are made."[31]

This surely merits a more in-depth explanation, as do the two main factors involved in debates over the bioethical dimension of respect for one's autonomy: the first is the establishment of the full mental capacity of the patient; the second the absence of external coercion. In the cases explored in the previous section, it is clear that neither of these fundamental features is present in Hunger Strikes, but rather the opposite.

[29]Sevinç (2008, pp. 655–659).

[30]It should be noted that in order to ensure the "successful" preservation at all costs of the lives of prisoners, Turkish authorities, through the utilisation of the Army, ended up producing a shocking number of casualties during the process of force-feeding. See, for example: http://news.bbc.co.uk/2/hi/europe/1739041.stm.

[31]Ho (2008, pp. 193–207).

As rightly pointed out by Sheila McLean in her book *Autonomy, Consent and the Law*,[32] in relation to the first point, it should be considered that, the unique, single-bodied notion of competence itself, has been questioned by numerous authors. Among these, Eike-Henner Kluge[33] affirms that there are as many as three different version of competence: cognitive, emotional and valuational.

While the first of these would overlap with the standards of mental capacity based on rationality, the latter two also take into account the value of an individual's wishes. In the Ben Garci case -as in most hunger strike cases- the aim was to protest, not to die. In some sense, then, we could affirm that both emotional and also valuational competence were absent in that situation, raising the suspicion that something more should have been done, possibly even with regard to the very acceptance of competence as the final way of deciding when it becomes acceptable for the state to intervene.[34]

Even the first version of competence could potentially be called into question, but the focus of this chapter will not involve a deeper assessment of mental capacity in this instance. It is necessary, however, to bear in mind that doubts remain over the level competence involved in a hunger strike.

Even more important is the acknowledgement that the absence of external coercion -the second fundamental condition indispensable to respect for autonomy- is inherently impossible in cases of hunger strike. In fact, be it for political or personal reasons, hunger strike is a form of protest that *requires* reference to other individuals; that is, to external factors independent of the striker. It follows that the presence of the -requisite- freedom from external coercion becomes doubtful, and with it the legitimacy of not having intervened before it was too late for Ben Garci.

Before further analysing this idea, other hunger strike cases will be taken into account in the next section.

4.6 Further Hunger Strike Cases

Among the numerous Hunger Strikes recently dealt with by different authorities around the world, three cases deserve particular attention in relation to our current investigation. Clearly each of them could be the subject of an entire standalone work, but on this occasion they will be discussed only schematically so as to provide the essential information specific to each of the cases.

---

[32]McLean (2010, p. 19).

[33]Kluge (2005, pp. 295–304).

[34]Notice that this is also relevant for cases of Anorexia Nervosa, as patients suffering from it do not want to die, they just prioritize the importance of their body image over death.

The first case is that of Orlando Zapata Tamayo,[35] a Cuban dissident who died on 23 February 2010 after 85 days of hunger strike. His intention was to protest against prison conditions, but -given that his accusations included "scorn of Fidel Castro"— it is not hard to believe that the authorities in charge had no intention of preserving his life, so they strategically decided to respect his will in the extreme. The similarities with the Ben Garci case seem obvious here, and acknowledging such a similitude with a non-democratic regime underlines even more uncompromisingly the weight of biopolitics in both cases and the rationale for questioning the Italian authorities' approach to the Ben Garci's hunger strike.

The second case is that of Irom Sharmila Chanu, in Manipur, India.[36] She has been on hunger strike -but it would probably more appropriate to say Death Fast in her case- for over 10 years as a form of protest against the Armed Forces Special Powers Act (AFSPA). In her case the authorities do not want her to die because they would prefer not to see her martyred, and for this reason she has been force-fed against her will for almost the entire duration of her hunger strike. The irony is that -apart from the process of force-feeding itself- Sharmila has suffered numerous forms of violence, including sexual abuse, as part of a campaign of intimidation aimed at persuading her to discontinue her protest. Yet she is not allowed to die. In this instance the authorities in charge decided to "hit the enemy" by keeping her alive rather than letting her die. This, as we shall see below, is a very common feature of biopolitics.

The third and final case comes back to the European context- Ukraine, more specifically- by examining disputes over the acceptable level to which authorities should adhere when choosing to force-feed a prisoner. In the Nevmerzhitsky case,[37] in fact, the European Convention on Human Rights (ECHR) affirmed that the regime of force-feeding being applied amounted in that context to torture. One interpretation of this might be that the ECHR was affirming that there is a limit beyond which the preservation of life cannot be considered a sufficient justification for enforcing treatment. Further, this could suggest that even in a context where the sanctity of life is the guiding principle, the realisation of the preservation of life might represent a greater moral wrong on the part of the authorities than letting the prisoner die. This case is interesting because it shows the ambiguity surrounding Hunger Strikes, especially in EU contexts. It also shows the practical relevance of Steps 2 and 3 considered in Sect. 4.4.

In the next chapter, we will look from close on the Guantanamo Bay, and that is why that example is not considered here. Having expanded on the biopolitical use

---

[35]See, for instance: http://edition.cnn.com/2010/WORLD/americas/02/23/cuba.dissident/index.html, http://news.bbc.co.uk/2/hi/8540162.stm [accessed on 4 January 2015].

[36]See, for example: http://news.bbc.co.uk/2/hi/south_asia/5348414.stm, http://www.dnaindia.com/opinion/comment_irom-sharmila-s-10-year-fast-is-ignored_1323405 [accessed on 4 January 2015].

[37]Nevmerzhitsky v Ukraine (2005) 43 EHRR 32 (ECtHR). Available at: http://sim.law.uu.nl/SIM/CaseLaw/hof.nsf/d0cd2c2c444d8d94c12567c2002de990/34464f8568936e2ec1256fd900316fca?OpenDocument [accessed on 4 January 2015].

and misuse of Hunger Strikes around the world, in the next section of this chapter we shall return to a more specific focus on the Ben Garci case.

4.7 A Biopolitical Distinction

In Ben Garci's case the biopolitical approach that led to non-intervention was based on three factors: Ben Garci was a prisoner, a foreigner and also a Muslim. According to a possible interpretation of Foucault,[38] these factors could function well as sufficient reasons for discouraging any active intervention by society, as the authorities in charge had no biopolitical intention to preserve his life. On the contrary, Ben Garci's life was a loss that could serve to send out three messages deriving from Ben Garci's specific condition: the "retributivist", the "precautionary" and the "religious" messages. Each will now be examined in turn.

First, let us unpack the "retributivist" message. As seen with the Singleton case, this is the most general of the three, working against criminals. If you have committed a crime -or the state finds that you did- there will be no exceptions, and you will pay for your crime to the fullest. It is in the state's interest -even if now considered to represent an outdated mode of governing- to communicate to the masses of potential criminals the inflexibility and incontrovertibility of the law. Aside from this repressive and more visible reason, however, there is an additional reading that would see a need on the part of society to deal uncompromisingly with the culpability of the criminal in order to be able to counteract it and prevent the spread of the "illness of crime". After all, as Foucault writes: "the criminal [...] represents a disease of the social body".[39] By ensuring his or her *just* suffering, society as whole can create the necessary antibodies to avoid infection.

Turning to the "precautionary" message, this level of the communication concerns migration and foreignness. The proposed biopolitical analysis of the state's decision not to intervene would suggest that the death of Ben Garci was used as a way of discouraging other foreigners from entering the country if not in the privileged position of having arranged stable employment before reaching Italy. This is a very sensitive topic in all Western countries, though illegal immigration is a particularly pressing issue in contemporary Italian politics. It is possible, then, that the Italian government did not want to lose the opportunity to exploit a self-inflicted death as a means of propagating a message of differential treatment for foreigners. In contrast with the quotation above, in fact, we should remember that, when talking about Marie-Antoinette in France, Foucault underlines that: "she is basically a foreigner, that is to say, she is not part of the social body."[40] Here then, there is a swing back from the previous position. The authorities are not interested in the

[38]Foucault (2010).

[39]Foucault, ibid., p. 91.

[40]Foucault, ibid., p. 97.

therapeutic solution of the problem of illegal immigration, for no ready answer is available to what is a global question that cannot be tackled by one state alone. The easiest way to deal with it such as not to damage the authorities' reputation before the electorate would be to provide a very strict response to the issue. Those who do not hold an Italian passport are not part of the Italian social body, and so, as a result of the combination of this notion with the considerations above, they are not to be cured within the system but rather prevented from joining it -especially if carrying the illness of criminality with them.

Finally, the "religious" message relates to the strong presence of the Vatican as a recognised player in the Italian political arena. This results in the need for political powers to take into account the Roman Catholic stance in certain cases in order to avoid undermining consensus and losing votes, particularly in relation to bioethical issues. Given that the Vatican decided not to intervene in an attempt to save Ben Garci's life -unlike in other recently disputed cases centred on the notion of the sanctity of life-[41] it is reasonable to sustain that such differential treatment of two human beings could be traced back to their religious affiliation.

In relation to these points, it would be interesting to take into account what John Williams highlights in his article "Hunger-Strikes: A Prisoner's Right or a 'Wicked Folly'?".[42] In the delicate discussions that have been taking place on both sides of the Atlantic -to focus on the Anglo-American context- it has been concluded in many situations that an explicit acknowledgement of the fact that state interests can legitimately counter-balance the acceptance of Hunger Strikes.[43] This factor underlines that the role of state authorities in this process is indeed political. This begs the question as to which variables can impact upon the evaluation of a specific change in policy towards a particular hunger strike. Once again, Foucault's work provides valuable insights into this problem: "It seems to me that essentially there have been only two major models for the control of individuals in the West: one is the exclusion of the lepers and the other is the model of the plague victims."[44]

This affirmation allows for an interesting application of its claims to the present investigation, especially if combined with a second Foucauldian notion that

[41]The widely publicised case of Eluana Englaro is probably the best example in recent years. See, amongst others: Englaro, B. and Nave, E., Eluana—La Libertà e la Vita, Rizzoli, 2008; http://news.bbc.co.uk/2/hi/europe/7880070.stm [accessed on 4 January 2015]; Bock et al. (2007, pp. 1041–1042); http://news.bbc.co.uk/2/hi/6186347.stm [accessed on 4 January 2015].

[42]Williams (2001, pp. 285–296).

[43]See, amongst other cases: Commonwealth of Pennsylvania, Department of Public Welfare, Fairview State Hospital v Joseph Kallinger (1990) 134 Pa Cmwlth 415, 580 A 2d 887). Despite being recognised to be competent -and thus entitled to refuse nutrition- the Court felt that Kallinger was trying to use the system to his advantage and therefore the Court opted for enforcing treatment on him for the sake of State interests.

[44]Foucault, ibid., p. 44.

"society responds to pathological criminality in two ways or offers a homogeneous response with two poles: one expiatory and the other therapeutic."[45]

When applied to our analysis thus far, we could say that religion -or at least the political bodies that represent the material aspects of a spiritual/religious message- works in the same way. From a conservative religious perspective -referring in particular to Roman Catholicism in the Italian context- the others (those not following the same credo) are eternally wrong and thus, by not converting themselves to message of the only true God, they need to atone for their sins in this life. Of course, this attitude has become much milder in recent decades, but the argument here emphasises the deep-rooted logic of expiation that was behind the non-action of not saving a life. One therapy could certainly have been conversion, and, perhaps if Ben Garci had decided to convert, developments in authorities' approach to the case would have been very different.

4.8 Conclusion

In the course of this chapter, we have explored one case that might be considered a cornerstone of American bioethics, namely that of Dax Cowart. We studied once again questions regarding the objective applicability of the notion of autonomy in that context, affirming that -even if perhaps the best choice was made- the justification for enforced treatment lacked both moral and legal grounds within the current framework that allows a competent patient to refuse treatment. In order to stress further the inconsistency, we then contrasted the Dax case with the recent fatal hunger strike of Sami Mbarka Ben Garci in Italy. To understand the complexity of the issues at stake in Hunger Strikes, a deeper analysis of the principles involved in its acceptance and permissibility was carried out, juxtaposing these principles the notion of sanctity of life and the moral duty of authorities to preserve it at all costs.

Through the use of different hunger strike cases a broader perspective on the topic has been presented, stressing the differential application that the same principles have had in otherwise similar circumstances.

The argument of this chapter affirms that the reason for a differential prioritisation of certain principles over others in relatively similar contexts stems from the biopolitical value of each case. As a result, while the preservation of life would be the main priority of authorities in order to avoid the martyrdom of an individual (with all the destabilising implications that such an action would produce), in Ben Garci's case the principle of autonomy was given primacy in deciding how to proceed in that context. The unconvincing supremacy of autonomy over sanctity of life (a particularly unusual value-hierarchy in Italy) leaves more than a little doubt over the authenticity of the authorities' decision. In addition, the rushed attempt to

[45]Foucault, ibid., p. 34.

save Ben Garci's life at the last moment, was symptomatic an inconsistent approach towards respecting the patient's desire or lack thereof to be rescued from certain death.

If future cases similar to that of Ben Garci or Dax are to escape the biopolitical reading that accuses the relevant authorities of discriminatory behaviour, we should aim to arrive once and for all at standard positions with respect to Hunger Strikes, Death Fasts and forced treatment in general.

Throughout this chapter additional controversies related to the use of the notion of autonomy in bioethical and legal contexts have been highlighted. Especially towards the end of the chapter, a biopolitical reading of the facts has been foregrounded, affirming that certain distinctions in bioethics are in fact functional to their political value. The next chapter of the book will look to provide a deeper engagement with the biopolitical dimension linking all of the cases considered in this work through a close look at a specific case: Guantanamo.

References

Areen, J., et al. 1984. *Law, science and medicine*. Mineola: Foundation Press.

Bock, M., V. Ciarrocchi, and C.J. Wiedermann. Case involving end-of-life decision issues in Italy. *Intensive Care Med* 33.

Brockman, B. 1999. Food refusal in prisoners: A communication or a method of self-killing? The role of the psychiatrist and resulting ethical challenges. *Journal of Medical Ethics* 25.

Burt, R. 1979. *Taking care of strangers*. Free Press.

Childress, J.F., and C.C. Campbell. 1989. Who is a doctor to decide whether a person LiVES or dies? Reflections on Dax's case. In *Dax's case-essays in medical ethics and human meaning*, ed. by L.D. Kliever. Dallas: Southern Methodist University Press.

Dworkin, R. 1993. *Life's dominion*. London: Harper Collins

Engelhardt, H.T., Jr. 1989. The limits of personal autonomy: The case of Donald 'Dax' Cowart. In *Dax's case-essays in medical ethics and human meaning*, ed. by L. D. Kliever. Dallas: Southern Methodist University Press.

Foster, C. 2009. *Choosing life, choosing death-the Tyranny of autonomy in medical ethics and law*. Oxford: Hart Publishing.

Foucault, M. 2010. *Abnormal: Lectures at the Collège de France 1974–1975*. Navayana Publishing.

Garasic, M.D., and C. Foster. 2012. When autonomy kills: The case of Sami Mbarka Ben Garci. *Medicine and Law* 31(4): 589–598.

Gillon, R. 2003. Ethics needs principles -four can encompass the rest- and respect for autonomy should be 'first among equals'. *Journal of Medical Ethics* 29.

Harris, J. 1994. *The value of life*. London: Routledge.

Ho, A. 2008. The individualist model of autonomy and the challenge of disability. *Bioethical Inquiry* 5.

Jonsen, A. 2003. *The birth of bioethics*. Oxford: Oxford University Press.

Kalk, W.J., M. Felix, E.R. Snoey, and Y. Veriawa. 1993. Voluntary total fasting in political prisoners: Clinical and biochemical observations, *South African Medical Journal* 83.

Kenny, M.A., D.M. Silove, and Z. Steel. 2004. Force Feeding Hunger striking asylum seekers. *Medical Journal of Australia* 180: 5.

Kluge, E.H. 2005. Competence, capacity and informed consent: beyond the cognitive-competence model. *Canadian Journal on Aging* 24.

Koçan, G., and A. Öncű. 2006. From the morality of living to the morality of dying: hunger strikes in Turkish prisons. *Citizenship Studies* 10(3).
McLean, S.A.M. 2010. *Autonomy, consent and the law*. Routledge-Cavendish.
O'Neill, O. 2002. *Autonomy and trust in bioethics*. Cambridge: Cambridge University Press.
Sevinç, M. 2008. Hunger strikes in Turkey. *Human Rights Quarterly* 30.
White, R.B. 1989. A memoir: Dax's case twelve years later. In *Dax's case-essays in medical ethics and human meaning*, ed. by L.D. Kliever. Dallas: Southern Methodist University Press.
Williams, J. 2001. Hunger-strikes: a prisoner's right or a 'Wicked Folly'? *The Howard Journal of Criminal Justice* 40(3).
Zaner, R.M. 1989. Failed or ongoing dialogues? In *Dax's case-essays in medical ethics and human meaning*, ed. by L.D. Kliever. Dallas: Southern Methodist University Press.

Chapter 5
Guantanamo and Its Specific Biopolitical Charge

5.1 Introduction

After having looked at a number of controversial Hunger Strikes from all over the globe, in this final part of the book the analysis will be devoted to possibly the most [in]famous detention centre on the planet: Guantánamo Bay (Guantanamo). There are three main reasons for which the specificity of Hunger Strikes and enforced medical treatment in Guantanamo deserve a separate and conclusive chapter: first, its symbolic power is evident, demanding for an engagement (critical or supportive) from anyone willing to deepen their understanding of the problems concerning Hunger Strikes, force-feeding and enforced medical treatment. Second, the way in which the US exercise their sovereignty over Guantanamo (due to its geographical and political specific status within US law) is unique and worthy of consideration. Third, the political impact that this detention centre has had—and still has—on the dialectics of the political world (Western and non-Western alike), represents a very good example of the biopolitical dynamics that are in place in cases of enforced medical treatment. In fact, as highlighted with the cases considered in the previous chapters, this practice affects the political world as much as the bioethical arena. By analysing from up close this specific case, this conclusive chapter aims at underlining how—within the liberal framework dominating Western society—a rigid juxtaposition of the sanctity of life and the principle of individual autonomy fails to take sufficiently into account the asymmetry of power relations in place in cases of Hunger Strikes—epitomised by Guantanamo.

M.D. Garasic, *Guantanamo and Other Cases of Enforced Medical Treatment*,
SpringerBriefs in Ethics, DOI 10.1007/978-3-319-22653-8_5

5.2 Declaration of Malta on Hunger Strikes

With the Declaration of Malta, the World Medical Association[1] has attempted to provide some clear guidelines for medical personnel involved—more or less directly—with the practice of Hunger Strikes. Without listing all the points in the document, some aspects need to be highlighted for their specific relevance to a comprehensive analysis of the situation in Guantanamo.

In the preamble of the document, it is clearly stressed how doctors must take into account (in all cases where some kind of coercion is in place) that despite the presence of competence, those prisoners have a restricted version of freedom in front of them. In addition (although not openly acknowledged), the Declaration of Malta specifically suggests a political engagement if certain practices do not comply with the accepted norm. The first principle in the document states that there is a "duty to act ethically. All physicians are bound by medical ethics in their professional contact with *vulnerable* people, even when not providing therapy. Whatever their role, physicians must try to prevent coercion or maltreatment of detainees and *must protest* if it occurs."[2]

As the first (and thus presumably the most important) principle of the document, these sentences contain a number of problematic terms that need attention. To begin with, we are introduced to a group of people defined as "vulnerable" as if this should simplify our understanding and categorisation of who is vulnerable. However, this is not a convincing move, and I shall briefly explain why.[3] To support my statement, it would be useful to take into account that, in 2013, the UNESCO International Bioethics Committee published a *Report on the Principle of Respect for Human Vulnerability and Personal Integrity*,[4] in which specific distinctions between vulnerable groups were made. Aside from invoking other key definitions to legitimise the need of implementing "vulnerability" in bioethical debates that are not meant to be covered by the present investigation,[5] the UNESCO International Bioethics Committee notably writes: "While some groups of people can always be considered vulnerable because of their status (e.g. children), others

---

[1]World Medical Association. Declaration of Malta on Hunger Strikers. Op. Cit.

[2]World Medical Association. Declaration of Malta on Hunger Strikers, Op. Cit. My emphasis.

[3]It is not the intention of this work to criticise the notion of vulnerability per se. Works such as Robert Goodin's influential book *Protecting the Vulnerable: A Reanalysis of Our Social Responsibilities* for example, are to be considered valuable and praiseworthy enterprises that have contributed (and can still contribute) to an improvement of the human condition and the progress of humanity, but doubts remain over the use of the notion made here. While in Goodin's account vulnerability is used also as a tool to criticise a too individualistic version of autonomy, in the UNESCO International Bioethics Committee report, vulnerability is used to reinforce the polarised conceptualisation of the *sanctity of life* versus *individual autonomy* dualism that this chapter refutes. Goodin (1985).

[4]UNESCO (2013).

[5]Human dignity and human rights being the main two.

may be vulnerable in one situation but not in another. Therefore, vulnerability cannot be considered as a one-off concept."[6]

If vulnerability is *not* a one-off concept then, applying this report to the Declaration of Malta brings forward the existence of different categories of vulnerability. Contextualising this awareness to situations of Hunger Strikes, the implementation of vulnerability does not seem to help clarifying who needs to be helped (and why) or not. That is so, because—among other variables—the conceptualisation of hunger strikers implicitly presupposes some kind of vulnerability (why would the person strike otherwise?) that can or cannot be legitimised by the contingent political agenda.[7]

The second term that requires some analysis is that of "protest". Physicians are obliged (and expected) to protest if they see some unethical behaviour towards the detainees. Yet, this "easy-fix" statement does not seem to bring much clarification on how we are to understand such a definition, and once again we need to further explore what does it mean in practice. First of all, who should be expected to hear Guantanamo physicians' protest? Authorities one would think, but who would those be in the given context? The force-feeding of a prisoner would not take place without the green light received from the military and political authorities in the first place. The conscious physicians could of course decide to protest publicly (jeopardising in some sense the legitimacy of their political leaders and institutions in the eyes of the international community), but this is a different choice from that of wanting to respect the detainee—and a much more political one than emphasised in the document. In addition, the "obligation" to protest could push the physician to take action in a morally questionable way for both him and the detainee. For example, while an ethical physician would be able to take over from the unethical colleague in treating the prisoners (guaranteeing the highest level of humanitarian behaviour within the set schemes of a force-feeding procedure), once labelled as a "traitor" of the code of conduct of the operation, she would be pushed aside and restrained from dealing with the detainees further, leaving them in the hands of less and less compassionate colleagues.

These considerations are linked to a number of other principles and definitions of the Declaration of Malta. In particular, the second, third and fourth principles give relevance to respecting individual autonomy, balancing benefits, harms and dual

[6]UNESCO International Bioethics Committee, *Report on the Principle of Respect for Human Vulnerability and Personal Integrity*, Op. Cit.

[7]Exploring the concept of vulnerability goes beyond the scope of this work, but a quote from Yechiel Michael Barilan nicely captures the essence of my point here: *"In technologically advanced societies, all people, but especially those who depend on others, are vulnerable to subtle or even unintended forms of exploitation and harm. Religions, social, and cultural practices whose goal is the protection of the vulnerable are also sources of vulnerability and abuse—especially of women and gay people. Numerous children have been sexually molested by clergymen who abused religious authority and infrastructure. The advent of nuclear power and mounting concerns about possible environmental disasters have prompted scientists and ethicists to underscore the vulnerability of the human race and its supportive ecosystem as a whole. The discipline of bioethics was born out of such concerns."* Barilan (2012, p. 141).

loyalties in ways that—albeit brief—require a short commentary. I will get back to the issue later in the chapter, but it is important to stress from the beginning of our analysis that prioritizing respect of individual autonomy over the sanctity of life is a dichotomy that is, in itself, charged with a number of a priori conditions that do not belong to the sphere of medicine. This important point is also linked to the way dual loyalties are portrayed. The problem with this way of conceptualising the situation is very well described by Dani Filc, Haddas Ziv, Mithal Nassar and Nadav Davidovitch in their article on Hunger Strikes in Israel:

> Emphasizing hunger-strikes as political acts implies that health practitioners' conducts cannot be analysed only through the prism of clinical medical ethics. It requires a public health ethics' approach, an approach that understands health in the context of the complex relationship between society, state, organizations, communities and individuals. [We present] an alternative to the liberal framing of the ethical dilemma as one in which the individual physician must choose between respecting the prisoner's individual autonomy and the sanctity of life.[8]

In line with this analysis, even benefits and harms (portrayed by the Declaration of Malta as the only possible result of the neat polarisation between competent and incompetent people) could be read in a different manner from that provided by the Declaration of Malta. For example, if "avoiding 'harm' means […] not forcing treatment upon competent people",[9] we would have to take this assessment of "who is competent" to be fixed in time. In other words, once established the presence (or absence) of competence in an individual, we would be able to act in accordance in a consistent manner throughout time. However, it has already been highlighted in the previous chapters that it is not so, and I will further the point in Sect. 5.3. In addition, if we combine this consideration with the statement in the preamble of the Declaration of Malta of the limited version of freedom under which competence is —supposedly at least—present, the overall picture becomes more blurry than perhaps envisaged by the World Medical Association. Before going back to a more theoretical analysis of the concepts at stake, the next section will highlight the specificity of Guantanamo in practical terms.

5.3 How Do Hunger Strikes in Guantanamo Bay Differ from Others?

Thanks to a perpetual lease agreement, Guantanamo (a 45 square mile area of Cuba), the US were granted the right to use the area for coaling and naval operations in 1903. In addition, the document expressly states that although the US exercise complete jurisdiction and control over and within such areas, Cuba retains the "ultimate sovereignty". As a result, since 1903, Guantanamo has been used as a

[8] Filc et al. (2014, p. 230).

[9] World Medical Association. Declaration of Malta on Hunger Strikers, Op. Cit.

US naval base, restricting its area to private use and limiting access and navigation without US authorization.

After an initial use directed only towards military purposes, Guantanamo was turned into a detention camp for asylum seekers in the US during the early 90 s. However, it was only in January 2002 that Guantanamo started to get the attention of the international community. The media started to report its name as the US transferred into the detention camp hundreds of individuals taken as prisoners in the military operations in Afghanistan started by the Bush administration as a response to the 9/11 attacks. These people were captured and held in Guantanamo without charge but instead defined as "unlawful combatants". This unprecedented definition (tailored specifically by the US Department of Defense so as to suit the emergency status wanted by the Bush administration), provided the detainers with unorthodox powers and threatening opportunities that—as was very recently officially admitted by the US authorities-[10] led to a misuse of such exceptional status for the interrogators. Article 16 of the 1984 Convention against Torture and Other Cruel, Inhuman or Degrading Treatment or Punishment[11] further explains why Guantanamo was chosen as a detention and interrogation centre for suspects of terrorism after 9/11. Being defined as outside US jurisdiction by the Bush administration for years (to the extent that the same administration affirmed that the prisoner could not have been granted any of the protections of the *Geneva Conventions*), this location constituted an ideal context in which to apply controversial "enhanced interrogation" techniques such as waterboarding. The specific advantage of having such interrogations in Guantanamo was that even if torture was proven to have taken place (the Bush administration always denied any allegation, but President Obama has recently specifically referred to those episodes as cases of torture[12] and the recent CIA report stressed it even further), the US government would have authorised its personnel to use such a treatment on a prisoner in a territory not under its jurisdiction, ensuring that a technical breach of the Convention against Torture and Other Cruel, Inhuman or Degrading Treatment or Punishment had not taken place. Eventually in 2004, the US Supreme Court determined that the US courts had jurisdiction, and it subsequently ruled in the 2006 *Hamdan v. Rumsfeld* case,[13] that prisoners had to be granted the minimal protections listed under Common Article 3 of the *Geneva Conventions*. As of 7 July 2006, the Department of Defense issued an internal memo affirming that the personnel working in Guantanamo would be required to comply with the directives, ensuring to protect detainees under Common Article 3 in the future.

With such a controversial development, it cannot be a surprise that—from the initiation of Guantanamo as a detention centre for potential terrorists—plenty has

[10]CIA (2014).

[11]UN (1984).

[12]I discuss more specifically the definition of torture and its link with Guantanamo elsewhere: Garasic (2015).

[13]*Hamdan* v *Rumsfeld* (2004).

been written on its exceptionality from a legal and moral point of view. Already in 1996, legal scholars defined Guantanamo as an "anomalous zone",[14] while after the beginning of the War on Terror wanted by the Bush Administration, prominent philosophers such as Judith Butler[15] and Giorgio Agamben[16] forcefully criticised the legal exceptionality of the status of the detainees in Guantanamo as a political move aimed at guaranteeing only the 'bare life'[17] of the prisoners by placing them outside of the rule of law—with the foreseeable result of a gradual dissolution of that life we usually refer to as "normal". Inspired by Carl Schmitt's work[18]—who developed an important relation between exception and authority—Fleur Johns[19] has replied to such critiques, arguing that we should instead see such an exceptionality of Guantanamo as a framework from within the law. It is only by defining it in those terms that we are able to fully understand "the scope for political action within such a juristic zone."[20]

For the purpose of the present investigation, both these positions can be considered valid because—whichever standpoint one might have—Guantanamo represents a unique example of widespread enforced medical treatment that provides valuable inputs into the discussion on the legitimacy of such a practice.

Before proceeding further, one important aspect about the present investigation should be clarified. In giving relevance to how Guantanamo's prisoners are in an exceptional position outside the legal sphere usually granted to individuals in the democratic world, we should not forget that their status of anomaly is not relevant for the assessment of the [absence of] medico-ethical conduct by the physicians involved with the practice. Legally and politically, Guantanamo's prisoners might indeed be a unique case, but medically speaking they remain equally entitled patients as any other detainees. Be it in Guantanamo or anywhere else in the world, the professional conduct of a physician is expected to be consistent—at least theoretically.

It is also important to understand that in this—evidently limited—analysis of the events in Guantanamo, it is absolutely not the intention of this work to legitimise, support or defend the practice of torture that took place in Guantanamo. The [exc] use of a political cause to justify one's medical malpractice has already had shady examples in the past century and does not need to be reanalysed in depth in order to be condemned, given that—as mentioned in the beginning of this book—they provided the main input in the creation of a number of medico-ethical documents that constitute the core of the code of conduct of doctors in the West after WWII. I will expand on the issue in the next section.

---

[14]Neuman (1996, p. 1197).

[15]Butler (2004, pp. 50–100).

[16]Agamben and Raulff (2004).

[17]Ibid.

[18]Schmitt (1996).

[19]Johns (2005, pp. 613–635).

[20]Ibid, p. 635.

5.4 The Role of Doctors and Their Dual Loyalty

As mentioned already in Sect. 5.1, the balancing of dual loyalties is one of the principles of the Declaration of Malta and—not surprisingly—it is one of the most controversial themes in the debates surrounding Guantanamo. As a result, two recent articles in the *New England Journal of Medicine* discussed the moral implications for doctors involved in situations of Hunger Strikes –particularly in relation to Guantanamo's prisoners. They call on physicians to refrain from force-feeding inmates for two reasons: (1) force-feeding is a direct violation of one's autonomy, and complicity of physicians in the act implies violation of medical ethics,[21] and (2) if physicians indeed stop force-feeding, the conditions of at least some prisoners will improve.[22] Here, a few additional considerations concerning the issues at stake are worthy of attention. In what follows, I will briefly review these papers and the ethical dilemmas raised by them.

George Annas, Sondra Crosby and Leonard Glantz[23] discuss the role of doctors involved in the enforced medical treatment of prisoners in Guantanamo, affirming that those participating in the "preservation" of the hunger strikers, are acting unethically. The main reason behind such an affirmation is based on the claim that —given the level of competence present in the hunger strikers—doctors should respect the prisoners' wishes. Of course, being an active part of the force-feeding procedure that involves some kind of intentional parallel harm[24] is hardly in line with the Hippocratic oath.[25] On the contrary, cases like that of the *Nevmerzhitsky v Ukraine*[26] have proven that sustaining the life of the individual in itself is no guarantee of humane treatment. In fact, in that case it was clearly demonstrated that enforced medical treatment can be considered torture.[27]

However, two clarifications must be made: first, as mentioned already in Chap. 4,[28] hunger strikers can—and often do—lose their competence directly because of the strike. The fact that they engage in the strike with full cognitive capacities does

[21]Annas et al. (2013, pp. 101–103).

[22]Gross (2013, pp. 103–105).

[23]Annas, G., Sondra S. Crosby, M.D., and Leonard H. Glantz, Op. Cit.

[24]See for example: Guantanamo: Medical Ethics Free Zone (2013); Leopold (2013); Moqbel (2013).

[25]Jotterand (2005, pp. 107–128).

[26]Nevmerzhitsky v Ukraine, Op. Cit. This has been a landmark case in which the European Court of Human Rights condemned the Ukrainian Government for having breached a number of articles of the European Convention on Human Rights. In particular, it was established that force-feeding a prisoner can be considered torture.

[27]This case had a particularly strong political connotation that makes the assessment of torture very evident, but in other instances the involvement of doctors aiming at not abandoning a patient gives rise to a medical dilemma that we cannot cover here. For an insightful analysis of this dimension of the problem see: Lepora and Millum (2011, pp. 235–254).

[28]See footnote 24 in Chap. 4.

not necessarily mean that they will maintain the full range of these capacities throughout the strike.[29]

The second qualification relates to the second reason for considering force-feeding as torture: the patient's right not to be harmed. Force-feeding is often perceived to be a painful and degrading act, serving as another method of enhanced interrogations. I agree that force-feeding could be conducted in an extremely painful and uncomfortable manner, and in this case should be considered torture. However, one has to bear in mind that there can be other ways of force-feeding, without the need to impact so dramatically on the prisoners' psyche. Aside from the nocturnal treatment of anorexic patients[30] considered in Chap. 2, naso-gastric treatment is also commonly used in cases of cystic fibrosis, where parents often choose to feed their children while asleep to reduce the pain and psychological burden of the procedure. As a result, this method is gaining popularity even among adults suffering from the same condition.[31] If the intention was exclusively to preserve the lives of prisoners, more benign methods of feeding could have been used. Instead force-feeding has been used as a political tool in Guantanamo.[32]

This very political dimension of Guantanamo is emphasized by Michael Gross.[33] He considers more directly the overlapping of the political and medical spheres, rightly pointing out that while the authorities might be able to accommodate the prisoners' requests in some cases, in other instances this is not possible because the requests are beyond acceptable levels –at least in terms of practicality, not of what the prisoners deserve morally. For example, prisoners could ask for a bigger cell, and authorities might grant certain improvements in the conditions of the inmate, but if their request directly conflicts with the existing authority in charge (i.e. demanding its dismantlement as a result of not recognising such an authority) the issue becomes much more political—making it impossible for the people in charge to give into the specific request. This deeply problematic aspect is further compounded by the legal complexity inherent in Hunger Strikes, that put hunger strikers in a very specific position in the eyes of the law. Whether we accept or not the force-feeding as something morally permissible is a complex ethical dilemma and it

---

[29]Fessler (2003, pp. 243–247).

[30]See footnote 27 and 33 in Chap. 2.

[31]Smith (2008, pp. 257–262).

[32]Of course, for all these techniques (naso-gastric treatment or percutaneous endoscopic gastrostomy for instance) to not be considered torture, the individual would have to consent in being fed in the first place, else he or she would remove the tube by their own means. If she or he shall be prevented from doing so, limitation of movement, coercion and force would be required. Thus, those methods of feeding could also be considered as torture. However, my point is that the level of burden of such enforced medical treatment would be much less invasive in psychological terms, and—to an extent—more in line with the claim that the applied coercion is in truth used *only* for the patient's best interest.

[33]Gross, M., L., Op. Cit.

is a prime example of a case in which descriptive ethics could and should inform normative ethics.[34] If at all allowed, it should be practiced under strict legal scrutiny with a valid way of assessing its implementation within a legal framework that is accepted as fair and just. What must be clear is this: any well functioning legal system has to assume to be just in the vast majority of cases (otherwise its very legitimacy is at risk because it would imply a structural injustice intrinsic in the system), but exceptions to the rule might exist. It is crucial to evaluate how the legal system can make such exceptions without becoming blackmailable. Otherwise, Hunger Strikes might expose an unfairness of treatment among prisoners serving different sentences, or between hunger strikers and non-hunger strikers.[35]

As highlighted in the previous chapter, Hunger Strikes are the quintessential representation of biopolitics: the State can decide whether to let one die or keep one alive in accordance with what is most functional to its political message. This crucial awareness of the dynamics in place should help us understanding the multi-layer complexity that Hunger Strikes carry with them: they do not only represent a medical or bioethical issue, but they are—by default—also and always a political act, thus to be considered a [bio]political issue as well. This aspect emphasizes another important dimension of Hunger Strikes and death fasts. These phenomena represent in many contexts (certainly Guantanamo is one of those) the only available "tool" with which prisoners attempt to send out a political message. Not allowing for their continuation in the name of the respect for the hunger strikers' lives can thus be seen as an efficient way for the authorities to mitigate the political effects of the strike. Aside from a political interference, in terms of medical ethics, this is a limiting factor for an individual's right to choose competently according to his own best interests.

As already stressed in Sect. 5.1., the prioritization of autonomy over the sanctity of life or other ethical parameters (such as the moral and professional duty for a physician to assist someone who is dying for example) is not as a politically neutral step as portrayed by supporters of the idea that autonomy is the leading principle to be respected in medical ethics. The relevance of this consideration will be developed in the following sections.

---

[34]Sugarman and Sulmasy (2010a).

[35]There are numerous scenarios in which a consistent consideration of the pleas of the hunger strikers is not feasible. For instance: if I was to hunger strike to have my time in prison reduced by half, would it change if I had been sentenced to 30 or 5 years? Surely it would in strict mathematical terms: 15 against 2 and a half years of reduced prison time. And if a group starts hunger striking to have their prison cells' conditions improved, how should authorities respond? It seems logical (and morally sound) that those conditions could be improved *only* by including also the non-hunger strikers –hence implicitly admitting the legitimacy of the protest in the first place.

5.5 Arguments Supporting the Use of Naso-Gastric Treatment in Guantanamo

Although the use of naso-gastric treatment clashes directly with the directives of the Declaration of Malta, here I want to speculate over the possibility to morally justify the use of naso-gastric treatment in Guantanamo. Aside from a thought experiment, this section will help us unveil some of the limits of the Declaration of Malta in addressing the problem of Hunger Strikes.

The a priori justification used to legitimise the disparity of [enforced] medical treatment in relatively similar cases is based on the assumption that authorities are capable of grasping a standard and objective version of autonomy and applying it fairly in specific instances.

For example, the distinction between the autonomous action and the autonomous person, as explained by Alasdair Maclean in his book *Autonomy, Informed Consent and Medical Law*, is the result of one specific way of understanding autonomy that contrasts the "objective truth" of what is autonomous in itself with the subjective capability of potentially—but not necessarily—perceiving it. He writes:

> Autonomous persons will not always act autonomously and, where they do not, the act may be contrary to their long-term autonomy or other interests. This raises the thorny, but crucial, question of whether it is more harmful to interfere with a present non-autonomous act or to allow that person to harm his or her autonomous life or future autonomy. Furthermore, the choice between protecting any decision of an autonomous person and only those decisions that are themselves autonomous has implications for the law since the latter position would justify a significantly greater degree of interference.[36]

There is no doubt that the latter position results in greater interference by the authorities, but what is particularly attractive and merit-worthy in this idea is the fact that it manages to underline the necessary relevance of the political dimension of what can be permitted to an individual.

The contrast between the possibility of respecting an autonomous person's decisions "no matter what" and the scenario in which his or her decisions could not be allowed due to their contravention of the law flags up an important incongruity. The definition of an action as autonomous does not depend upon the outcomes of the action—as claimed in the second sentence of the quotation above—but instead upon the a priori legal permissibility. This way of understanding the law is certainly peculiar considering the well-accepted distinction between the *actus reus* and *mens rea* ordinarily applied in Anglo-American contexts and already explained in some depth in Chap. 3.

Moreover, there exists a certain tension between what is affirmed by Maclean in his first sentence and what can be deduced from the rest of the quotation. Surely, if a person goes to jail for she robbed a bank, this will affect her interests and future autonomy; but, if we consider her guilty of that crime and decide to put her in jail,

[36]Maclean (2009, p. 12).

we must accept that—as well as the *actus reus*—the *mens rea* was also present. Otherwise—as already outlined in previous parts of the book—we could not judge her to be guilty, at least not entirely.

This is the paradox: in the scheme provided by Maclean, *mens rea* has the potential to always be present, therefore the suggested solution for complicated cases seems to be the prevention of the *actus reus*.

However, this way of dealing with the issue appears self-evidently unconvincing. If we stop people from acting on the grounds of what they are thinking, and potentially planning to do, we would enter a vortex that would quickly lead to the end of democracy and, more specifically, the dismantling of current legal systems that depend on the presumption of innocence. In fact, were such a power to be established as legitimate, it would be relatively easy to affirm that a person was "just about to do something" and preventively incarcerate them.

In other words, adapting this consideration to the terminology used in the cases studied in this book, one might be defined as an autonomous agent in terms of competence/capacity and yet not be allowed to choose freely certain options not deemed correct by the authorities. In such a situation, however, doubts emerge over who defines the "correct".

The impossibility of detecting the "real" outcomes of an action for an agent is what undermines the second dimension of autonomy described above. Certainly, the dimension more directly related to freedom is this one, as it raises more questions regarding the influence of politics in establishing the correct way of understanding the specificities of a particular person's case in order to deem him a qualified citizen entitled to be an autonomous person capable also of following a prescribed definition of autonomous action. However, assessing matters that concern how to live one's life is a particularly challenging task in a society such as our Western one, where respect for individual choice constitutes such a fundamental value within our worldview.

In his book *Life's Dominion*, Ronald Dworkin[37] stresses the difference between the personal value of life and its intrinsic value, with the intention of highlighting that personal value is the result not only of the foreseeable success and achievements that an individual can hope to experience within his lifetime, but also of its mistakes and sorrows. The natural (or divine) and human "investments" of which he talks about in order to reconcile conservatives and liberals (at least momentarily) represent a scale upon which to measure many kinds of intrinsic readjustment that are active in reshaping our autonomy (or, more accurately, competence or freedom depending on the context). It follows, thus, that an individual's understanding and affirmation of his autonomy can vary significantly in relation to physical circumstances such as, say, a serious sports injury.

In fact, as detailed in Chap. 1, Dworkin's vision of autonomy is centred on the notion that we should evaluate the past actions of an individual, his choices, desires

[37]Dworkin, R., Op. Cit.

and directives, as these could prove helpful in assessing the level of (what he terms) authenticity.

In the same way that a physical accident might result in a decisive growth for an individual, so might a decisional accident. If we preventively disbar a human being from making even wrong decisions (within reasonable bounds) for the sake of a pre-established understanding of autonomy, we would not only deny her freedom of choice, but, in utilitarian terms, we would also deprive the wider humanity of the valuable contributions that each one of us can have by "investing" in himself or herself.

If we add that, the same Declaration of Malta states in its second principle that "hunger strikers' true wishes may not be as clear as they appear",[38] there might be room for accepting naso-gastric treatment in Guantanamo as we might be not so sure that this is *not* what they want. Of course, this does not have to entail purposely violent and humiliating methods. As explained in the course of this work,[39] naso-gastric treatment can be implemented in ways that are less violent and invasive that normally associated with the practice. For example, feeding the patient while asleep would prevent most of the psychological burden of having a medical procedure forced upon oneself. However, no argument can be made to support torture and this issue will be addressed more in depth in the next section.

5.6 Arguments Condemning the Use of Naso-Gastric Treatment in Guantanamo

Mark Mercurio[40] flags up the well-accepted practice, during the medical training period, of allowing not yet fully competent physicians to operate on patients as part of their training. Could we argue that this way of intending and conceptualising consent in fact conflicts with Kantian ethics?[41] Arguably, Kant's universal law clashes with the practical truth that some patients are "forced" to accept treatment by doctors less capable than others, or, quite simply, not as well-trained as others, accepting thus their own use as means to the end of producing expert doctors.

As Kant famously stated: "Act in such a way that you treat humanity, whether in your own person or in the person of another, always at the same time as an end and never simply as a means."[42]

This idea, widely assumed to underpin the Western tradition of medical practice, merits greater attention: we as a society accept a system that, by default, allows its members to be used as means to certain ends. However, it seems that we are still

---

[38]World Medical Association. Declaration of Malta on Hunger Strikers, Op. Cit.

[39]See footnote 31 above.

[40]Mercurio (2008, pp. 44–57).

[41]Le Marvan and Stock (2005, p. 514).

[42]Kant., I., Grounding for the Metaphysics of Morals, Op. Cit., p. 36.

incapable of recognising such generalised contravention of the Kantian Laws, while we instead focus our attention on specific cases where the idea of using an individual as a means to an end is deemed unacceptable. This inconsistency in analysing different situations reveals the political issues behind current debates over autonomy in medical ethics. The acceptance of such an inconsistency in medical practice is based on the idea of defending "the patient's best interest", but, as affirmed by Mercurio himself,[43] this view is rooted in a utilitarian model rather than on one willing to preserve a Kantian notion of autonomy. Thus, if we are to have a linear approach, we must either abandon the use of autonomy as the principle under which we reject naso-gastric treatment (exposing the utilitarian nature of our stand), or reconnect more deeply to a more coherent account of autonomy. This issue will not expanded further here, but it is important to note that—aside from the "standard" arguments used against enforced medical treatment (including the Declaration of Malta) based on the need to respect one's autonomous choice—the most powerful reason for condemning the use of naso-gastric treatment in Guantanamo is its torturous nature.[44] As detainees on hunger strike have been force-fed in the most horrific and humiliating ways,[45] it appears clear that the situation was used by the [medical] personnel involved to apply yet another degrading treatment to Guantanamo's prisoners.

This is certainly not the first time that political authorities have decided to force-feed someone in order to reduce their political impact,[46] but it is possibly the first time in which the torturing nature of the procedure has been camouflaged as medically valid. If we are not to completely lose faith in both our physicians and our politicians, we cannot tolerate such a stand.

---

[43]Mercurio, M., R., Op. Cit, p. 49.

[44]Prominent scholars such as Peter Singer and Frances Kamm have argued that torture can be morally justified under certain (rare) circumstances. However, speculating over the acceptability of such a stand is not relevant for the present investigation, because if that argument were to be applied to Guantanamo the whole idea of force-feeding the detainees in their best interest would become redundant, as we would be enforcing medical treatment solely for the purpose of withdrawing sensitive information. Singer (2010); Kamm (2011).

[45]"As described in the context of the rectal feeding of al-Nashiri, [food] was infused into al-Nashiri 'in a forward-facing position (Trendlenberg) with head lower than torso.' [...] Majid Khan's 'lunch's tray', consisting of hummus, pasta with sauce, nuts, and raisins was 'pureed' and rectally infused." CIA, Committee Study of the Central Intelligence Agency's Detention and Interrogation Program, Op. Cit, p. 100.

[46]The force-feeding of suffragettes in Britain and France is a widely known for example. See amongst others: Miller (2009).

5.7 Hippocratic Oath or Political Agenda? A Biopolitical Analysis of the Issue

There is a general consensus in the field of bioethics that sees this discipline—or branch of philosophy—as separate from other fields that hold concepts such as freedom or autonomy at their centre; namely, political contexts.

In his book the *Birth of Bioethics*, when trying to establish whether or not bioethics can be described as a discipline, Albert Jonsen points out an important aspect peculiar to this field: "Only half of bioethics counts as an ordinary academic discipline […] The other half of bioethics is the public discourse: people of all sorts and professions talking and arguing about bioethical questions."[47]

This affirmation, of course, speaks to the intrinsic mission of bioethics, aiming—at least—to allow a direct interaction with the non-expert, with normal people, the masses.

If we accept such a mission for bioethics, we must simultaneously become aware of the fact that a disappointment in the bioethical sphere can have direct repercussions in the public sphere (including the medical sphere), and it is for this reason that we urgently need to find a way to avoid such disappointments.

In line with this "separatist" approach, for example, in *Pragmatic Bioethics*, Gleen McGee[48] argues that this detachment from its humanist heritage defines the specific character of bioethics: an application of theories in specific—in their vast majority medical—cases. He views this detachment as a positive development.

However, I do not share this view, for the reason that, although bioethics can be defined as a separate discipline with a particular way of dealing with ever-changing practical issues rather than more essentially philosophical ones (such as injustice or causation, for example), its core values are strongly intertwined with the surrounding disciplines (law, politics, medicine) and must coherently and convincingly apply certain directives to distinct contexts. Otherwise, we risk not only a loss of trust in the field of bioethics, but also in its related fields, including politics.

In order to properly understand the unevenness of applications of the notion of autonomy in relatively similar cases, I believe—in open contradiction of McGee's approach—that the answer has to be found in the bigger picture rather than through limiting our analysis to a specific set of cases. As an example of the misleading way of understanding autonomy in bioethics, one final important aspect is highlighted by James Stacey Taylor. He writes: "The feature of practical autonomy that is at issue here is its (supposed) content-neutrality."[49]

This distinction is crucial when we make the a priori assumption—as often is the case in bioethics—that the kind of autonomy to which we refer to in bioethical contexts is different from that applied to politics. As such, I shall affirm that—in line

[47]Jonsen (1998, p. 346).

[48]McGee (2003). This view is emphasised particularly strongly in the introduction, but it is present throughout the book.

[49]Taylor (2009, p. 64).

with Taylor—autonomy is not content-free and that content-neutrality does not exist.

As it has been described, in more "scientific" contexts autonomy translates into competence or capacity where these concepts are considered to be more "objective" ways of assessing the capability of the individual to perform a certain task or action and, in doing so, to show their autonomy.

However, through the use of the various examples put forward in the previous chapters, it has been shown that the concepts of competence and capacity are not insulated from "political directives" and are hence subjective. It follows that the very interpretation at which we arrive as a society—with certain common values, principles and shared dynamics shaping our perception of the acceptable—is not content-neutral but, rather, content-biased.

The biopolitical impact of Guantanamo hunger strikers is specific because they have been characterised as a defined group of people by the very fact that they are confined there. This differs from other instances of Hunger Strikes where the political dissident already had a "clashing identity" in the eyes of the authorities in charge of the detention centre (i.e. criminal records, belonging to a different ethnic or religious group). Yet, for better or worse, the interaction that physicians can and do have with the detainees does not differ from other contexts. This awareness does not guarantee an advancement in understanding how a physician should behave in absolute terms, but it helps in contextualising Guantanamo better.

5.8 Conclusion

The deeply contrasting way in which Hunger Strikes are dealt with globally allude to their role in the biopolitical agenda, and their intrinsic political nature is not sufficiently stressed by the Declaration of Malta. The centrality of individual autonomy (arbitrarily prioritized over the sanctity of life) does not provide a content-neutral reading of the situations, but it instead reinforces a predefined way of conceptualising the situation—resulting into an unhelpful moral and professional code of conduct for all those physicians facing cases of force-feeding. To expect the US government to escape such dynamics is worthy of consideration and respect, but it is perhaps too naïve. Society cannot expect doctors to necessarily disagree with the political program of a certain government because of their specific role. However, doctors too (be it in Guantanamo or elsewhere), cannot hide behind Eichmann's attitude: "I was just doing my job", because that is an insufficient reason to justify a potentially torturous practice. If they decide to take a side in what might be considered a political stand, they could be entitled to do so, but only if they are willing to accept the burden that such a choice implies.

References

Agamben, G., and U. Raulff. 2004. Interview with Giorgio Agamben—life, a work of art without an author: the state of exception, the administration of disorder and private life. *German Law Journal* 5. http://www.germanlawjournal.com/index.php?pageID=11&artID=437. Accessed 4 Jan 2015.

Annas, G., S. Sondra, M.D. Crosby, and L.H. Glantz. 2013. Guantanamo Bay: a medical ethics–free zone? *The New England Journal of Medicine* 369(2): 101–103.

Barilan, Y.M. 2012. *Human dignity, human rights, and responsibility*. Cambridge: MIT Press.

Butler, J. 2004. Indefinite detention. In *Precarious life. The powers of mourning and violence*, ed. J. Butler, 50–100. London: Verso.

CIA. 2014. Committee study of the central intelligence agency's detention and interrogation program. http://www.fas.org/irp/congress/2014_rpt/ssci-rdi.pdf. Accessed 4 Jan 2015.

Fessler, D.M.T. 2003. The implications of starvation induced psychological changes for the ethical treatment of hunger strikers. *Journal of Medical Ethics* 29(4):243–247.

Filc, D., H. Haddas Ziv, M. Nassar, and N. Davidovitch. 2014. Palestinian prisoners' hunger-strikes in Israeli prisons: beyond the dual-loyalty dilemma in medical practice and patient care. *Public Health Ethics* 7(3): 229–238.

Garasic, M.D. 2015. Torture. In *Encyclopedia of global bioethics*. New York: Springer. doi:10.1007/978-3-319-05544-2_420-1.

Goodin, R. 1985. *Protecting the vulnerable: a reanalysis of our social responsibilities*. Chicago: University of Chicago Press.

Gross, M.L. 2013. Force-feeding, autonomy, and the public interest. *The New England Journal of Medicine* 369(2): 103–105.

Guantanamo: Medical Ethics Free Zone. 2013. Russia Today. http://www.youtube.com/watch?v=Y77OOY3UJ1Y. Accessed 4 Jan 2015.

Hamdan v Rumsfeld. 2004. U.S. Dist. LEXIS 22724, at 21–27.

Johns, F. 2005. Guantanamo Bay and the annihilation of the exception. *The European Journal of International Law* 16(4): 613–635.

Jonsen, A. 1998. *The birth of bioethics*, 346. New York: Oxford University Press.

Jotterand, F. 2005. The Hippocratic oath and contemporary medicine: dialectic between past ideals and present reality? *Journal of Medicine and Philosophy* 30: 107–128.

Kamm, F.M. 2011. *Ethics for enemies: terror, torture and war*. Oxford: Oxford University Press.

Le Marvan, P., and B. Stock. 2005. Medical learning curves and the Kantian ideal. *Journal of Medical Ethics* 31: 514.

Leopold, J. 2013. Revised Guantanamo force-feed policy exposed. Al Jazeera. http://www.aljazeera.com/humanrights/2013/05/201358152317954140.html.

Lepora, C., and J. Millum. 2011. The tortured patient: a medical dilemma. *Asian Bioethics Review* 3(3): 235–254.

Maclean, A. 2009. *Autonomy, informed consent and medical law*, 12. Cambridge: Cambridge University Press.

McGee, G. (ed.). 2003. *Pragmatic bioethics*. Cambridge: MIT press.

Mercurio, M.R. 2008. *An analysis of candidate ethical justifications for allowing inexperienced physicians-in-training to perform invasive procedures*. *Journal of Medicine and Philosophy* 33: 44–57.

Miller, I. 2009. Necessary torture? vivisection, suffragette force-feeding, and responses to scientific medicine in Britain c. 1870–1920. *Journal of the History of Medicine and Allied Sciences*. doi:10.1093/jhmas/jrp008.

Moqbel, S.N.A.H. 2013. Gitmo is killing me. New York Times. http://www.nytimes.com/2013/04/15/opinion/hunger-striking-at-guantanamo-bay.html?_r=0. Accessed 4 Jan 2015.

Neuman, G.L. 1996. Anomalous zones. *Stanford Law Review* 48:1128–1233.

Schmitt, C. 1996. *The Concept of the Political.* Trans. T.B. Strong. Chicago: University of Chicago Press.

Singer, P. 2010. *The life you can save: acting now to end world poverty*. Melbourne : Text Publishing Company.

Smith, D.L. 2008. A nocturnal nasogastric feeding programme in cystic fibrosis adults. *Journal of Human Nutrition and Dietetics* 7(4): 257–262.

Sugarman, J., and D.P. Sulmasy. 2010a. The many methods of medical ethics (or, thirteen ways of looking at a blackbird. In *Methods in medical ethics,* ed. J. Sugarman, and D.P. Sulmasy. Washington: Georgetown University Press.

Taylor, J.S. 2009. *Practical autonomy in bioethics*, p. 64. New York: Routledge.

UN. 1984. Convention against torture and other cruel, inhuman or degrading treatment or punishment. http://www.ohchr.org/EN/ProfessionalInterest/Pages/CAT.aspx. Accessed 4 Jan 2015.

UNESCO. 2013. International bioethics committee. Report on the principle of respect for human vulnerability and personal integrity. http://unesdoc.unesco.org/images/0021/002194/219494E.pdf. Accessed 4 Jan 2015.

Chapter 6
Conclusion

In conclusion, it appears clear that the line between being forced to be medically treated and being allowed to refuse medical treatment is indeed very thin. This already minimal distinction becomes even fuzzier when we bring into the picture the interdependent terms of freedom, consent and autonomy, themselves used interchangeably by various actors in bioethical and political contexts.

Informed consent -a notion introduced to Western bioethics with the successful aim of enhancing the rights of the patient vis-à-vis enforced treatment- presupposes a level of competence that will allow the individual to process the information provided and subsequently to make an autonomous decision regarding her treatment.

Competence (or capacity, technically speaking) is the term used to denote our level of autonomy, a concept near-impossible to grasp in a objective, "scientific" way. With the rise of this concept, patients were guaranteed more power over decisions concerning their health, and thus their freedom. In fact, a person has come to be allowed to exercise her freedom by autonomously choosing what to do with her life as long as she is deemed capable of making certain decisions, comply with certain tasks, and so forth. However, despite the undeniable improvements that this innovation has brought about, this book has considered various cases where the supposed objectivity that is at play in the evaluation of an individual's level of competence is, in fact, deeply entrenched in politics.

In order to highlight the complexity of such interactions it was necessary to provide an analysis of the current role of autonomy in Western bioethics, developed at two parallel levels: on the one hand, there existed a need to foreground specific cases (with their unique problems) so as to give the critique a more "real-world" edge and to flag up inconsistencies in the application of autonomy in various Western contexts (US, UK and Italy), demonstrating the urgent need to find a more satisfactory way of dealing with similar cases in the future.

On the other hand, autonomy had to be considered in a broader sense that would not limit its analysis to the field of bioethics. Hence, in order to contextualise our critique of the current way of inconsistently applying the notion of autonomy, the first step taken by this work was to provide a general understanding of the historical development of autonomy within the Western tradition, as well as the standard approaches utilised to define it.

M.D. Garasic, *Guantanamo and Other Cases of Enforced Medical Treatment*,
SpringerBriefs in Ethics, DOI 10.1007/978-3-319-22653-8_6

The initial premise of the book was that the advent of bioethics created new challenges for the way in which we understand and deal with autonomy. Since its incursion into philosophical debates, autonomy has needed constant readjustments in line with discussions over its appropriate definition and precise meaning, limits and value. In this process, the Anglo-American bioethical community has increasingly defined autonomy as a self-standing notion insulated from its political context, capable of being applied in different ways to relatively similar cases.

This work has examined the validity of such an approach by bringing to the fore bioethical cases that clearly highlight the fictional status of the Anglo-American notion of autonomy in a vacuum. I have argued that we live in an era where everything is fast-paced. Unavoidably, this also affects the approach that doctors have towards their patients, and so the room for dialogue has often been reduced. Yet, it is important to realise that some medical practices are already failing to comply with the current standard Western approach in bioethics that entails an unwavering application of the individual-centred notion of autonomy.

Such exceptions prove once more the subjugated status of bioethics in relation to its political context. We need to take this opportunity to pause and reshape the notion of autonomy in bioethical contexts with the intention of producing positive effects also in the political arena.

In order to support this vision, the book followed a structure that, having broadly defined autonomy in bioethics and politics, began with the controversial adaptation of this concept to a specific case of refusal of life-saving medical treatment, before gradually expanding on doubts over the legitimacy of its [mis]use.

Hence, in Chap. 1, we described how autonomy is currently defined in Western bioethics by reference to the two mainstream authors central to discussions of autonomy: Kant and Mill. Subsequently, autonomy was considered through the eyes of more recent and contemporary positions so as to prepare the ground for a debate over the validity of certain interpretations in counter-position to others, and, most importantly of all, to analyse what might incline us to accept (if at all) its differential application in different contexts.

After this introduction to general understandings and applications of autonomy in political and bioethical contexts, in Chap. 2 the reader's attention was directed towards practical cases where controversy over how autonomy should be -and indeed is- applied to real life instances is most tangible. The focus on cases of refusal of naso-gastric treatment by patients suffering from Anorexia Nervosa constituted the first of these examples because it forces us to question the objective validity of certain presumptions that Western society tends to have regarding issues of this kind. Beginning with these controversial cases also helped us point out a very important aspect that needs to be considered if a convincing assessment of the definition of competence is to be achieved.

The standard argument against compliance with the refusal of naso-gastric treatment is based on the idea that the autonomy of an individual should always be preserved and respected as long as it is possible to affirm that the individual is sufficiently competent to be deemed autonomous. The validity of this tautological approach, criticised thoroughly by the work of Giordano, was called into question, pointing out that we cannot categorically deny that, even if incompetent in relation to food, anorexics can still be considered competent enough to make decisions regarding their quality of life. Taking this position as reasonable, the justification for not accepting a refusal of treatment on the grounds of respecting an individual's autonomy becomes rather weak. Indeed, it becomes weaker still if this hesitancy on the part of authorities involved in the decision turns out to be related to a well-defined political message concerning the necessary preservation of life itself.

To establish the veracity of such a critical reading of cases of Anorexia Nervosa, in Chap. 3. I introduced an additional case where enforced treatment was instead carried out in order to allow a killing to take place. The Singleton case extended the reach of discussions relating to the distinction between the theoretical realm and the practical question of what society should do in order to preserve and encourage respect for the autonomy of the individual, including in its practical implementation in current law. The biased application of categorisations of mental illness raised further questions over the function of defining certain individuals as competent or not in relation to how this limits their potential for action within a given system. In stark contrast with the previously explored cases of Anorexia Nervosa, in fact, Singleton was forcibly treated against his will in order to enable his execution.

The "preservation of life at all costs" attitude applied in cases of Anorexia Nervosa (based on the still dominant view in Western bioethics that the sanctity of life must be defended where possible), seemed to have dissipated so as to make room for the conflicting societal desire to ensure that the necessary continuity of retributivism would not be jeopardised by one exceptionally complicated case. In further analysing this disparity, it was suggested that the enforced treatment suffered by Singleton could not be -and should not have been- justified even in conventional medical terms, as it appears obvious that curing someone only to kill him can hardly be described as "acting in his best interest" -the standard Hippocratic approach in Western medicine.

Having explored the above situations as examples of a strategic use of autonomy in relation to cases involving mental illness, I then went on to consider other cases where the incompetence of the patients was not inherently implied in the condition of the person in question. In Chap. 4, the Dax case and the Ben Garci case brought to the fore some additional doubts over the real nature of uneven treatment on the part of the authorities in relatively similar bioethical cases. These doubts compounded concerns over the relationship between the misuse of autonomy in bioethical cases and their political dimensions.

Through an analysis of the unconvincing justification of enforced treatment (for Dax) and of its more or less complete absence (in the Ben Garci case), Chap. 4 underlined more vehemently the need to understand these cases not as standalone exceptions to a well-functioning rule, but rather as signs of reinforcement of a certain biopolitical agenda.

Once again the crucial distinction between the two cases was related to the contrast between the disparity of application of the same notion: while for Dax the assessment of temporary incompetence led to a prolonged forced treatment, Ben Garci's enforced treatment was deliberately delayed on the grounds that respect for his autonomy had to be ensured, despite the fact that Hunger Strikes have been proved to produce temporary incompetence towards their final stages. I argued that, contrary to what in fact occurred, this awareness should have prompted authorities to apply the same care for Ben Garci as they did in other circumstances (including Dax's), a change in approach which did not occur because of the political salience of the case.

In comparing the two cases towards the end of the chapter, I utilised the work of Foucault to provide a biopolitical reading of the distinctions made in the two instances, affirming once again that it is impossible to ignore the political weight that each of the cases had in its specific context.

In this respect it should be clear to the reader that -as explained in the preface- owing to the conscious choice of this work to utilise a number of sources to support the critical reading of autonomy that it hopes to have achieved, it has been necessary to limit the depth with which these various positions have been explored.

The Foucauldian analysis employed here proceeded along the same lines. To be fair towards both his work and mine, it is important to understand that although my application of his theory refers to the discussion over autonomy with the intention of destabilising one of its interpretations in favour of another, I am aware that a more complete coverage of Foucault's point of view would have required discussion of power relations even in the new scenario.

In other words, differently from this book's objective, Foucault's critical analysis cannot be satisfied by the idea of principled autonomy as a concept free of power schemes and dynamics. I do not wish to contest the validity of such a view here, but it is important to acknowledge that I am aware of its existence and relevance.

Finally, in Chap. 5 the critique was expanded through the analysis of the current situation in Guantanamo, stressing how -within the Western liberal tradition-contrasting the sanctity of life and the principle of individual autonomy in cases of hunger strike does not give sufficient relevance to the power dynamics behind the resulting actions taken by political and medical authorities in such a context. In relation to that, this chapter was functional to explore the limits of the Declaration of Malta and the problematic role that doctors have in cases of hunger strike. Most importantly however, the force-feeding of hunger strikers in the detention centres in Guantanamo represented a valuable contribution to this work for three reasons: the widely covered visibility of the case, the uniqueness of Guantanamo's detainees legal status and the impact (passed and potential) of this situation in the understanding and reshaping the limits of the notion of autonomy in biopolitical terms.

As reiterated throughout this work, the definition of autonomy within the current individual-centred system cannot provide reasonable grounds upon which to override refusal of treatment in cases of Anorexia Nervosa or Hunger Strikes (as preferable to the political and medical authorities), instead favouring enforced treatment in other contexts where the ultimate aim is to not preserve a life, such as the Singleton case.

This acknowledgement should not prevent us from seeking more consistent alternative applications of the notion of autonomy. On the contrary, it is through this increased understanding that we should realise that autonomy could be used in a more consistent way if we were to apply the non-individualistic variant of it suggested by Onora O'Neill, which incorporates notions of duty and obligation towards the community. This may represent the best way of redefining the current Western approach to autonomy without falling into an inconsistent application of it: we are free to choose individualistically as long as our choice does not significantly affect society (be it the family, the state or the community). Where it does affect society, authority comes into play and reaffirms what can and cannot be tolerated.

Obviously, this restriction of one's freedom in favour of authority is already present in the biopolitical dimension criticised throughout this work. My proposal for reducing the misuse of power is to enlarge the gamut of actors involved in decision-making so to ensure a more valuable form of dialogue, genuinely driven by the intention to serve the best interests of the person rather than a potentially malicious biopolitical agenda.

A practical outcome of this reform might be the rethinking of the role of authorities in relation to individual autonomy. The increase in respect for the patient's informed choice has undeniably been a positive achievement for the whole society, but it is perhaps time for us to state more clearly our need to reemphasise responsibility. The upshot of this would be the difficult acceptance of the legal implications of such an innovation in bioethics, which would in themselves mark the first reduction of the autonomy of the individual in the past 20 years. In fact, in relation to this point, we should not forget that Kant's and Mill's standard interpretations of autonomy would not leave much room for the permissibility of enforced treatment on a competent adult. As such, the person in question should be entitled to choose freely how to pursue her own understanding of happiness, which for anorexic patients might mean the end of the repetitive, demoralising procedure that naso-gastric treatment represents. The only way that a possible justification could be found -and this is indeed the method used in the current legislation to prevent the interruption of treatment- has to focus on Kant. That being the case, some of the less considered aspects of his view should be given greater relevance in order to rebalance the current bioethical crisis that we are faced with. Within the wide range of possible ways of reading Kant, the application of O'Neill's approach to cases of refusal of treatment in Anorexia Nervosa may be the only way that will finally allow us to escape the unsatisfactory application of Categorical Imperatives currently favoured by our society. If bioethics is willing to keep on using the term "autonomy" without inconsistency, substantial changes will have to be made to its interpretation.

To conclude, and even if unable to substantiate the practical outcomes of such a suggestion in this instance, I believe that, in more practical terms, Italy could represent an ideal state in which to implement this new less exclusively individual-centred version of autonomy.

If we consider the fact that the Italian context (unlike the Anglo-American tradition) has never made such a neat switch towards that unquestioned prioritisation of the individual-centred variant of autonomy described in this book, we could hardly disagree that this aspect (presupposing agreement on the validity and necessity of a change in approach regarding autonomy) has created the perfect conditions for Italy to represent an example of how Western bioethical debates could be reshaped.

Going back to the critique considered in Chap. 1, where Bruce Jennings argued against O'Neill's suggestion that a detachment from the individual-centred form of autonomy is deeply unrealistic given its deep embeddedness in American society, we could affirm that by the same logic Italian society represents a far more conducive context in which to put into practice the concept of principled autonomy. Given that Italy is a liberal country that has never quite embraced the variant of autonomy criticised in this work, it remains in the advantageous position of not being destined to follow the same path towards the structural injustices manifest in other contexts, injustices bound to degenerate if not properly tackled, as evidenced by the unacceptable, politically driven resolution of the Ben Garci case.

Bibliography

Bok, H. 2003. Freedom and practical reason. In *Free will*, ed. G. Watson. Oxford: Oxford University Press.
Dworkin, G. 1976. Paternalism. In *Moral problems in medicine*, ed. S. Gorovitz. Englewood Cliffs: Prentice-Hall.
Eisenberg, L. 2004. Medicating death row inmates so they qualify for execution. *Virtual Mentor* 6(9).
Engelhardt, T. 2003. The many faces of autonomy. *Health Care Annual* 9.
Englaro, B. and E. Nave. 2008. Eluana – La Libertà e la Vita, Rizzoli.
Garasic, M.D. 2013. The singleton case: enforcing medical treatment to put a person to death. *Medicine, Health Care and Philosophy* 16(4): 795–806.
Maeder, T. 1985. *Crime and madness.* New York: Harper and Row Publishers.
Singer, P. 1994. Rethinking life and death: the collapse of our traditional values. New York: St. Martin's Griffin.
Smith, A. 2003. *The wealth of nations.* New York: Bantam Classics.
Wicks, E. 2001.The right to refuse medical treatment under the European convention of human rights. *Medical Law Review* 9(1):17–40.

Legal Cases and Documents

Singleton v Norris, 319 F.3d 1018 (8th Cir. 2003) (Habeas-Competency).
Singleton v Arkansas, 124 S.Ct. 74 (2003) (Cert. Denied).
Washington v Harper, 494 US 210 (1990).
Penry v Lynaugh, 492 US 302 (1989).
State v Taylor, S83428 (1989).
Parsons v Alabama, 81 AL 577, So 854 AL (1886).
Sinclair v State of Mississippi, 132 So. 581 MS (1931).
State v Strasburg, 110 P. 1020 WA (1910).
Leland v Oregon, 343 US 790 OR (1952).
Durham v United States, 214 F.2d 862 (1954).
Code of Medical Ethics, American Medical Association, Opinion 2.06, Capital Punishment, July 1980.
Mental Capacity Act, 2005.
Mental Health Act, 2007.
Concern for Dying. 1985. *Dax's Case, videotape*. New York: Unicorn Media.
President's Commission on Ethical Problems in Medicine and. 1981. *Biomedical and Behavioral Research, Defining Death: A Report on the Medical, Legal, and Ethical Issues in the Definition of Death (Washington.* D.C.: U.S. Government Printing, Office.

M.D. Garasic, *Guantanamo and Other Cases of Enforced Medical Treatment*, SpringerBriefs in Ethics, DOI 10.1007/978-3-319-22653-8

"Please Let Me Die", transcript of videotape available in Areen, J., et. al., *Law, Science and Medicine*, Mineola, Foundation Press, 1984.

Declaration of Malta on Hunger Strikers (adopted by the 43rd World Medical Assembly in Malta, November 1991 and revised at the 44th World Medical Assembly in Marbella, Spain, November 1992). Geneva: WMA, 1992. The document has been revised again in the 57th General Assembly in Pilanesberg, South Africa, October 2006.

Internet resources

http://www.psychiatricethics.org.uk/ANwebreport/report.pdf
http://www.who.int/classifications/icd/en/
http://cpancf.com/eatingdisorders.asp
http://www.who.int/mediacentre/factsheets/fs220/en/
http://news.bbc.co.uk/2/hi/europe/1722075.stm
http://edition.cnn.com/2010/WORLD/americas/02/23/cuba.dissident/index.html
http://news.bbc.co.uk/2/hi/south_asia/5348414.stm
http://ndpr.nd.edu/
http://www.youtube.com/watch?v=Y77OOY3UJ1Y
http://www.aljazeera.com/humanrights/2013/05/201358152317954140.html
http://www.nytimes.com/2013/04/15/opinion/hunger-striking-at-guantanamo-bay.html?_r=0

MIX
Papier aus verantwortungsvollen Quellen
Paper from responsible sources
FSC® C105338

If you have any concerns about our products,
you can contact us on
ProductSafety@springernature.com

In case Publisher is established outside the EU,
the EU authorized representative is:
Springer Nature Customer Service Center GmbH
Europaplatz 3, 69115 Heidelberg, Germany

Printed by Libri Plureos GmbH
in Hamburg, Germany